FEAR AND FORTUNE

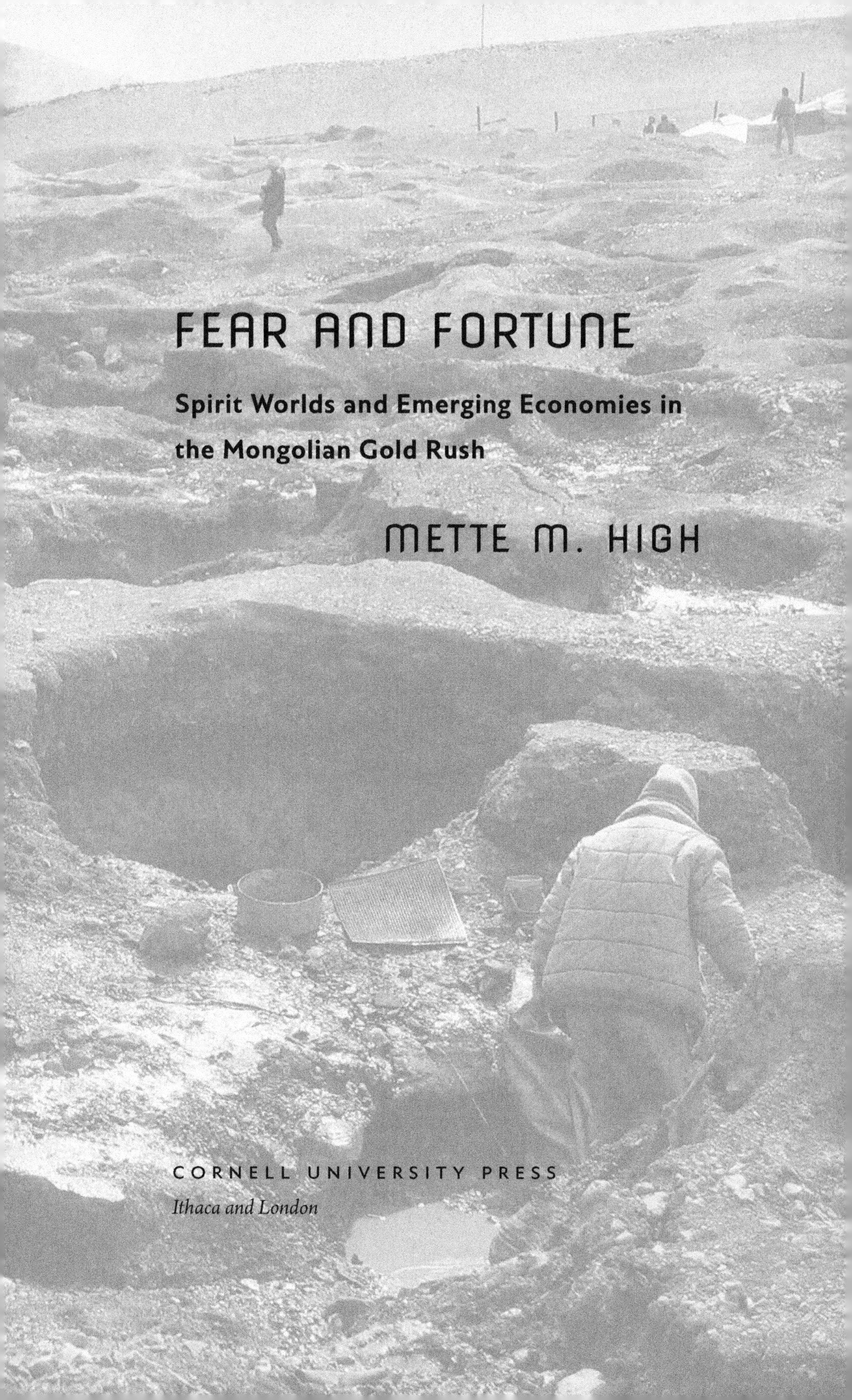

FEAR AND FORTUNE

Spirit Worlds and Emerging Economies in the Mongolian Gold Rush

METTE M. HIGH

CORNELL UNIVERSITY PRESS

Ithaca and London

First published 2017 by Cornell University Press
First printing, Cornell Paperbacks, 2017

Printed in the United States of America

Library of Congress Cataloging-in-Publication Data

Names: High, Mette M., author.
Title: Fear and fortune : spirit worlds and emerging economies in the Mongolian gold rush / Mette M. High.
Description: Ithaca : Cornell University Press, 2017. | Includes bibliographical references and index.
Identifiers: LCCN 2016047307 (print) | LCCN 2016050097 (ebook) | ISBN 9781501707544 (cloth :alk. paper) | ISBN 9781501707551 (pbk. : alk. paper) | ISBN 9781501708114 (epub/mobi) | ISBN 9781501708121 (pdf)
Subjects: LCSH: Gold mines and mining—Social aspects—Mongolia—Uyanga. | Economic development—Social aspects—Mongolia—Uyanga. | Gold—Religious aspects—Buddhism. | Ethnology—Mongolia—Uyanga.
Classification: LCC HD9536.M653 U933 2017 (print) | LCC HD9536.M653 (ebook) | DDC 333.8/541095173—dc23
LC record available at https://lccn.loc.gov/2016047307

Cornell University Press strives to use environmentally responsible suppliers and materials to the fullest extent possible in the publishing of its books. Such materials include vegetable-based, low-VOC inks and acid-free papers that are recycled, totally chlorine-free, or partly composed of nonwood fibers. For further information, visit our website at www.cornellpress.cornell.edu.

To my parents, Pia and Keld

Lately in a wreck of a Californian ship, one of the passengers fastened a belt about him with two hundred pounds of gold in it, with which he was found afterwards at the bottom. Now, as he was sinking—had he the gold? Or had the gold him?

John Ruskin, *Unto This Last,* 1862

To see what is in front of one's nose needs a constant struggle.

George Orwell, *Tribune,* March 22, 1946

CONTENTS

ILLUSTRATIONS

ACKNOWLEDGMENTS

This book has been a long time in the making and I have incurred many debts over the years. My deepest gratitude goes to the people of Uyanga, who so willingly shared their fears and fortunes with me. Their patience, humor, kindness, curiosity, and warmth made my fieldwork an incredible experience. I especially thank Baajiimaa, Tömörchödör, Ber, Yagaanövgön, and Ejee for taking such good care of me and supporting me through challenging times. I think about them a lot and cannot wait to see them again. I also thank Üjin, Erdemtögs, Düvshintögs, Alhaa, Rinchendorj, Davaasambuu, Bazarragchaa, Band, Nyambuu Aav, Degidsüren, Enhjargal, Ganbat, Bayasgalan, Battseteg, Baterdene, Amarjargal, Lhagva, Ulambayar, Tsegii, Horolgarav, Budlam, Nergüi, Sanchir, Tögslam, Banzragch, Battsüren, Pürevsüren, Buyanjargal, Nyamdolgor, Ganaa, Tsetsgee, Batzaya, Nergüi, Bilgee, Ganbaatar, Byamsüren, Choidogsüren, Bundrur, Odgerel, Dalai, Tsegii, Pürevtogtoh, Soylham, Genden, Mishigdorj, Lhagvadorj, Dolgosüren, and many others for so kindly allowing me into their lives. I have written this book out of my deepest respect for them.

My fieldwork in Uyanga was possible only as a result of the help and advice of numerous people in Mongolia. I thank Tümen Dashzeveg at the National University of Mongolia, the academic sponsor of my research in Mongolia, and Otgontugs, secretary in its international department, for helping me with my numerous visas. I also thank members of the Ongi River Movement, ILO-IPEC, the American Center for Mongolian Studies, and the Asia Foundation in Ulaanbaatar. I also thank my research assistants, Ama, Bulga, and above all, Boloroo, who has become a close friend and ensures I continue to stay up-to-date with the happenings in Mongolia. For introducing me to the world of geology and making sure, as he put it, that I not only knew about miners but also about mining, I am grateful to Robin Grayson. For their patient and generous exchanges about mining in Mongolia, I also thank Peter Appel, Batbuyan Batjav, Miles Light, Gantulga Mönh-Erdene, Bill Murray, Tümenbayar, and Tony Whitten. I am still impressed by the patience and optimism of my language teachers, Dogoo and Ogi. I also thank Elena, Garry and Sveta, Karsten and Else, Peter Marsh, Shijer, and Tom Sant for their warm hospitality and generosity during my visits to Ulaanbaatar. Trips across the Mongolian countryside with Enhee, Frank Wiederkehr, Christopher Hudak, and Vincent Galvin were simply great. Having experienced the challenges of fieldwork in Mongolia, Ann Benwell, Aude Michelet, Marissa Smith, and

Troy Sternberg provided valuable support and optimism. A surprise visit by Joe Long in the field was particularly memorable. Since my first visit to Mongolia, I have relied heavily on and am deeply grateful to my friends Momo and Nadia, who predicted that I would always return.

Since ninja mining is not a legal activity in Mongolia, I have sought to avoid revealing the faces of ninjas. I have therefore included only drawings and distant photos for illustration. I thank Jos Sances for his beautiful sketches and the map. I also thank Tim Franco for his photograph of a Buddhist ritual carried out at a Mongolian mining site and Jason Glavy for his kind sharing of the Soyombo digital font.

As an undergraduate student of anthropology, I made a trip to Cambridge and met with Caroline Humphrey to talk about Mongolia and mining. Ever since, I have been fortunate to benefit from her immense insights and support. She is a truly inspirational supervisor and mentor, whose work sets a standard for both regional and anthropological scholarship. In Cambridge, the Mongolia and Inner Asia Studies Unit provided an ideal research environment. I have drawn much from rigorous discussions at the local pub with Franck Billé, Ludek Broz, Uradyn Bulag, Bernard Charlier, Giovanni da Col, Grégory Delaplace, Hildegard Diemberger, Bumochir Dulam, Rebecca Empson, Signe Gundersen, Lars Højer, Chris Kaplonski, Morten Pedersen, David Sneath, Katie Swancutt, Olga Ulturgasheva, Hürelbaatar Ujeed, Rane Willerslev, and Astrid Zimmermann. I also owe much to Libby Peachey, who generously offered her calm, practical assistance. For support and friendship on and off the River Cam, I thank in particular Kelli Rudolph and Margaret Young.

The final form of this book has also benefited from questions, comments, and conversations from numerous other readers along the way. I thank Nicolas Argenti, Amit Desai, Florent Giehmann, Evan Killick, Nicolas E. Martin, Eleanor Peers, and Silvia De Zordo for keeping me company in the library. I treasured every cup of coffee and esoteric discussion they provided. I owe much to my colleagues and friends at University of St Andrews. In particular, I thank Sabine Hyland and Adam Reed, whose reflections on anthropology have been inspiring and their advice through various stages of the project invaluable. Sharing my work with my undergraduate and graduate students over the years has been particularly helpful. Their perceptive and constructive comments, along with their enthusiasm and excitement, have helped me see the significance of my material. For seminar invitations and great discussions, I thank Astrid Oberborbeck Andersen, Naomi Appleton, Judith Bovensiepen, Marc Brightman, Kate Browne, Catherine Dolan, Richard Fardon, Janne Flora, Martin Fotta, Eric Hirsch, Abby Kinchy, Mateusz Laszczkowski, James Maguire, Martin Mills, Mathijs Pelkmans, Christina Schwenkel, Jessica Smith, and Piers Vitebsky. For their inspiration and encouragement, I am grateful to Rita Astuti, Christopher Atwood, Debbora Battaglia, Laura Bear, Francesca Bray, Stephanie Bunn, Janet Carsten, Liana Chua, Tony Crook, Roy Dilley, Elizabeth Ferry, Stephan Feuchtwang, Chris Fuller, Peter Gow, Vanessa Grotti, Stephen Gudeman, Jane Guyer, Keith Hart, Casey High, Martin Holbraad, Stephen Hugh-Jones, Deborah James, Aimée Joyce, James Laidlaw, Michael Lambek, Jonathan Lear, Donald Lopez, June Nash, Peter Oakley, Jonathan

Parry, Nigel Rapport, Knut Rio, Katharina Schneider, Michael Scott, Jonathan Spencer, Marilyn Strathern, Jim Taylor, Christina Toren, Harry Walker, Vesna Wallace, Andrew Walsh, and Alexei Yurchak.

At Cornell University Press, I thank Roger Malcolm Haydon for his enthusiastic support and expert guidance in the development of this book. He understood my manuscript in a way that is any author's dream. I also thank Jamie Fuller, Karen Hwa, and Emily Powers for their help in preparing the book for publication, and I am grateful to the anonymous reviewers for the press for their careful readings and useful comments.

My research was facilitated by the generous funding of the following institutions: the Economic and Social Research Council, King's College Cambridge, the Cambridge European Trust, the Wenner-Gren Foundation, the Sigrid Rausing Fund, the Wyse Fund, and the Radcliffe-Brown Trust Fund of the Royal Anthropological Institute. A fellowship at the London School of Economics provided an exciting and rigorous environment in which I could begin conceptualizing this book. A residential writing fellowship at the Centro Incontri Umani, Switzerland, offered an ideal sanctuary in which I could begin writing it, and I am grateful to the Centro and its founder and director, Angela Hobart, for this immense support. The British Academy supported my continuing research and writing through a postdoctoral fellowship, and the Leverhulme Trust saw this book through to its end.

Introducing my family to life in Uyanga has been particularly important to me, and I am grateful that they have all been able to share in my experiences. My son, who made his first trip to Uyanga at the age of six months, loves nothing more than a good story from Mongolia, and he has asked me some of the most difficult questions about the gold rush. Their unfailing support, curiosity, and profound inspiration have been with me all the way, and I owe them much more than I can say.

Finally, I thank Taylor & Francis (www.tandfonline.com) for permission to use part of my article, coauthored with Jonathan Schlesinger, "Rulers and Rascals: The Politics of Gold Mining in Mongolian Qing History" from *Central Asian Survey* 29 (3) (2010): 289–304, in the introduction. I also thank John Wiley and Sons for permission to use part of my article "Polluted Money, Polluted Wealth: Emerging Regimes of Value in the Mongolian Gold Rush" from *American Ethnologist* 40 (4) (2013): 676–88, in chapter 4, and part of my article "Cosmologies of Freedom and Buddhist Self-Transformation in the Mongolian Gold Rush" from the *Journal of the Royal Anthropological Institute* 19 (4) (2013): 753–70, in chapter 5.

NOTE ON TRANSLITERATION AND TRANSLATION

There is no standard system for the transliteration of Mongolian words. This is partly because different scripts are concurrently used across the Mongolian cultural region. For example, the Republic of Mongolia draws mainly on the Cyrillic script, while Inner Mongolia still uses the classical vertical Uyghur-Mongolian script. The latter script conveys much more nuance in orthography and pronunciation than the Cyrillic script and has hence given rise to some transliteration systems that include many diacritical marks and non-Roman characters. Some of the other transliteration systems that refer to the Cyrillic script have on the contrary simplified the Mongolian words to such an extent that greater ambiguity and uncertainty in the original Mongolian spelling are introduced. In this book I follow Rozycki's (1996) scheme for transliteration with two alterations (Й as *i* instead of *y*, and Ы as *y* instead of *ih*). In rendering my transliteration close to conventional Halh (the largest ethnic group) pronunciation, I hope the reader will be able to get an immediate sense of the language.

In the case of relatively well-known Mongolian words, I have used the transliteration that is most commonly used in the English literature (such as Oyu Tolgoi, Ongi River, and khan). However, I have chosen to retain the Mongolian spelling of Chinggis Khaan rather than the English version of Genghis Khan since the romanization of the Mongolian spelling is becoming increasingly commonplace, both within and beyond Mongolia.

Table 1. Transliteration scheme used in in this book

А:	a	(c**a**r)	И:	i	(t**i**n)	Р:	r	(rolling "r")	Ш:	sh	(**sh**ow)
Б:	b	(**b**ike)	Й:	i	(t**i**n)	С:	s	(**s**afe)	Щ:	shch	(**shch**)
В:	v	(**v**ase)	К:	k	(**k**ey)	Т:	t	(**t**en)	Ъ:	"	(no sound)
Г:	g	(**g**old)	Л:	l	(bi**ll**y)	У:	u	(c**oa**t)	Ы:	y	(p**i**n)
Д:	d	(**d**ime)	М:	m	(**m**ap)	Ү:	ü	(p**u**t)	Ь:	'	("short" i)
Е:	ye/yö*	(**ye**arn)	Н:	n	(**n**ot)	Ф:	f	(**f**ine)	Э:	e	(p**e**n)
Ё:	yo	(**ya**cht)	О:	o	(h**o**t)	Х:	h	(lo**ch**)	Ю:	yu/yü*	(**you**)
Ж:	j	(**j**est)	Ө:	ö	(y**ea**rn)	Ц:	ts	(ca**ts**)	Я:	ya	(ka**ya**k)
З:	z	(a**dz**e)	П:	p	(**p**ie)	Ч:	ch	(**ch**at)			

*Disambiguation to reflect Mongolian vowel harmony.

With regard to the transliteration of Mongolian nouns in the plural form, I have often simply added an *s* to the singular form. This is because Mongolian plural suffixes can change the appearance of nouns significantly, and my intention has been to make it easier for the reader to recognize Mongolian words. The plural of *ninja* (informal-sector gold miner) is thus written *ninjas* rather than the Mongolian *ninja nar*. In order to shorten the Mongolian inserted text, I have also generally provided infinitive phrases except when the conjugated verb form was accompanied by important suffix chains. All translations, unless otherwise stated, are my own.

In the glossary I have included only those words that are used repeatedly throughout the book. All Mongolian words are accompanied by a translation and/or explanation in the text.

FEAR AND FORTUNE

Introduction

Land of Fortune

"I THOUGHT IT would be an opportunity to make some quick money," my friend Davaa said in a voice of regret on a late evening in 2001. He was in his midtwenties and had just returned from his first stint as a miner in Mongolia's "gold rush" (*altny hiirhel*, lit. excitement/hysteria for gold). Emptying his glass of vodka, he looked at me straight and added, "I had tried *everything* and this was my very last chance." Davaa glanced briefly at his fiancée, who sat quietly next to him. She had never really liked the idea of his becoming a so-called *ninja*—a colloquial term used by the miners themselves, the general population, and government officials to refer to "artisanal," or informal-sector, miners as opposed to those employed by mining companies. "Washing dirt" (*shoroo ugaah*) was hard, physical work that involved much more hazardous and grim working conditions than her own bank job in the capital city of Ulaanbaatar. Moreover, ninja mining was officially illegal, and the police were infamous for carrying out violent raids on mining camps. As it turned out, the dangers Davaa encountered in the mines were far worse than those his fiancée had initially feared. As she explained,

> When Davaa was in the mines, he heard that some people there are crazy. Apparently in the early evenings, when it starts to get dark, groups of miners scavenge the camps, searching for an able-bodied person who is alone. The miners carry knives, never guns, and when they find what they are looking for, they quickly surround the person. They first cover his head with a bag, so that no one can hear the person's screams. No one will notice what the miners are doing. They are quick and quiet. They then throw the person over their backs and carry him through the darkness up to the mountain where people make the usual offerings to spirits. But instead of offering milk or tea, the miners offer blood. Milk and tea are no longer strong enough for the spirits—they want more. They are not satisfied with dairy products, they require human blood. Once the miners arrive at the mountain, they carefully put their victim on the ground. They pull out their knives and stab the person again and again until streams of blood flow down the mountainside. That's why they have knives and not guns. Guns don't make as much blood, right? Have you ever heard of anyone who got shot in the mines? [I shook my head.] Well, that's why.

Davaa returned to Ulaanbaatar after only two weeks in the mines, yet many others are still drawn to the remote mountains of Mongolia in search of gold. For people like

Davaa, the gold rush, which has grown to become the largest ever on the Asian continent, involves major risks, perhaps even the sacrifice of human life itself. Although national and international commentators rejoice in Mongolia's immense mineral wealth, which is expected to help ease the global crisis in financial investment markets, gold is locally regarded as a volatile and inalienable material that is not readily exchangeable and commodifiable. In contrast to other kinds of metal, it is seen to retain strong ties to the landscape and its many spirit beings. Since these ties cannot easily be severed and are particularly strong at the point of extraction, the fortune of the precious metal is inseparable from the fears that surround mining. When Peter Munk—the founder and chairman of the world's largest gold mining corporation, Barrick Gold Corp.—commented that "gold finds are becoming more and more difficult," many Mongolians would certainly have agreed. But whereas Munk was concerned about the growing scarcity of gold, many Mongolians are concerned about the growing pursuit of gold in their country. As people try to capitalize on the vast and precious mineral wealth of Mongolia, fear and fortune go hand in hand.

This book is about the many thousands of people who take part in the Mongolian gold rush, concentrating on those most directly involved in the extraction and transaction of gold. It examines how herders, ninjas, Buddhist lamas, illegal gold traders, and other local traders experience what they themselves consider radical change or, as they put it, when life becomes strange (*hachin*) and chaotic (*zambaraagüi*). But rather than being about state protests, local resistance, or emancipatory opposition to a rapidly growing mining economy, it describes a process in which observable transformations are experienced as spiraling, disturbing, and not least unavoidable. The gold rush has arrived and it affects everyone—whether they want it or not.

Focusing on local transformations of a global gold economy, I explore how people make sense of the unprecedented extraction of gold and production of "gold money" (*altny möngö*), handled as potent objects that are inextricably linked to both human actions and spirit worlds. As the regional literature has shown, the production of wealth (*bayalag*) is related to people's specialized knowledge of a craft and their ability to attract the fortune or blessings (*hishig*) of local household and nature spirits (Chabros 1992; Empson 2011; High 2008a; Swancutt 2012). When people transgress taboos (*tseer*) by digging into the ground and panning the rivers for gold, spirit beings become upset and cause illnesses, accidents, and other misfortunes (see also Humphrey 1995; Gantulga 2011; Tseren 1996). Produced through immoral acts and involving the perils of pollution (*buzar*), gold money is not regarded as identical to money earned through other means. Whereas some financial forecasts present national currencies as seemingly uniform and reducible to comparable quantitative figures, people in the gold rush insist on the particularity and uniqueness of money earned from gold mining. Indicated physically by their mud and dirt, their wear and tear, the money notes are easy to identify. Drawing on and emphasizing this physical distinction, the various gold rush participants conceptualize and use gold money in very different ways. For some, it is considered heavy, prone to a dangerous stagnation that leads to misfortune. For others, it is considered lifeless and without a capacity for profit making unless it is brought into material contact with "renewed money" from afar. Circulating locally as an effectively devalued currency, gold money is considered part of a much-feared "cosmoeconomy" in which malicious gossip, misfortune, and pollution mediate human desires and angry spirits. Seeking the fortunes of gold is a socially and cosmologically dangerous, even if also materially rewarding, endeavor.

Whereas modernist paradigms postulate contradictions between economic practices and expressions of the "supernatural" in contemporary society, this book asks what happens when economic life is regarded as manifestations of spirit and other nonhuman phenomena. As Gabriel Tarde ([1895] 2012) pointed out long ago, the notion of "society" does not necessarily refer exclusively to human beings. Recognizing passions and potentialities to be innumerable and inexhaustible, interlaced and indeterminate, we cannot assume any domain of life to be populated by humans only. Accepting these limits to anthropocentrism, how can we best analytically approach growing involvements in transnational economies alongside a strong recognition of a society that is not exclusively, or even primarily, human? Rather than approaching economic life and spirit worlds as discrete and incongruent, this book focuses on their fundamental continuity and mutual composition. I demonstrate how the gold mines in the district of Uyanga in the Övörhangai province of central Mongolia have given rise to intense and unyielding economic, social, and spirit-religious ties. Given this convergence of local cosmologies and a feared resource-oriented economy, I suggest we need to pay more attention to the coexistence of multiple economies, beyond the boundaries of a nation-state and an exclusively human society. It is by also recognizing nonhumans as possibly being afforded the role of agents that we are in a position to understand the interests and concerns that underpin people's own views of change. This focus on how people experience radical transformation does not reflect a presumption that history has a beneficial direction, a homogenizing function, or a particularly capitalist imperative. Nor does it reflect the idea that its current order somehow lacks or evinces varying degrees of "authenticity." Rather, this book is centrally concerned with the creation of new ways of living at a time of cosmoeconomic upheaval.

Consequently, I am not drawing on the notion of "moral economy" in this book. Premised on the existence of a collectively shared morality that motivates specific economic practices and sentiments, this notion presumes a remarkable degree of consensus among its various participants. In Uyanga, if not also beyond, such a presumption is unwarranted. There is no such homogenous valorization with regard to how one "ought" to live economic life. Multiple beings coexist, and this coexistence is challenging rather than peaceful. Inspired by da Col's (2012a) insights into the interface between cosmology and economic life, I am therefore using the notion of cosmoeconomy here to elucidate the articulations between the actions, interactions, and transactions of various human and nonhuman beings. This notion is not intended as a critique of anthropological notions of the economy as somehow inherently unable to account for interconnected suprahuman or posthuman phenomena. It is also not intended as a heuristic invention that necessarily applies everywhere else in the world.

Rather, the notion of cosmoeconomy, which has been used for decades in other disciplines (Kogan 1992; Andersson et al. 1993; Gilder 2013), is intended to highlight connections and continuities where we, as analysts, might otherwise expect there to be none. Without imposing preconceived conclusions, it forces us to pay attention to the presence and importance of diverse agents. Drawing inspiration from people's own experiences of the Mongolian gold rush, I use it here as an analytical move that reminds us of the expansive and inclusive world within which people mobilize action.

Boomtown Mongolia

Mongolia has in recent years become a major and much-talked-about hub for global mining. It has vast reserves of mineral wealth, in particular gold, copper, and coal. A headline in the *Financial Times* (Hook 2011a) proclaimed the country to be "one of the world's last great mining frontiers, a freak of geology with more than $1,000bn in probable mineral deposits." Neighboring one of the world's biggest economies, Mongolia is conveniently located to supply what China needs to fuel its growth. Whether it is minerals for industrial production, domestic infrastructure, or strategic investment, the Mongolian region is plentiful. In September 2010, the country's then prime minister, Sühbaataryn Batbold, appeared on a popular US talk show to promote this wealth in natural resources. With a broad and excited smile, he announced, "There is huge potential! On top of [coal, copper and gold], we have new commodities to export to China. Iron ore, zinc. . . . And we also have prospects for oil and gas. . . . And important reserves of uranium!" Given Batbold's enthusiasm for the country's natural resource wealth, it is not surprising that by 2012 large parts of the country had been excavated and more than six thousand significant deposits were discovered. In that year, an astonishing 32 percent of the country was officially licensed for mining and exploration (personal communication, Mineral Resources Authority of Mongolia). The 3,865 licenses that were issued that year were not concentrated in a few isolated areas but scattered across all regions of the country.[1] In addition to the official licensing, there were also many illegal companies (*huul' bus kompanuud*) and thousands of gold rush miners who did not have licenses for the areas they mined. As the riches beneath the arid soil were increasingly revealed and exploited, the country's "freak geology" became an undeniably influential reality.

In 2011, when many countries were struggling with mounting debts, budget deficits, and economic slowdown, Mongolia experienced a 21 percent annual growth in its gross domestic product. It was heralded as "the country that is likely to grow faster than any other in the next decade" (*Economist* 2012). Its national currency, *tögrög*, received the international accolade of being the "best-performing currency in the world," making the *Financial Times* ask, "Kuwait 1950. Abu Dhabi 1970. Qatar 1995. Mongolia 2012?" (Pilling 2012). In a mining rush during which private equity houses and global investors were playing the role of prospectors, the Mongolian Stock Exchange was awarded the title of "the world's best performer," with share prices climbing 121 percent in 2010 (Hook 2011b). In the young and relatively unknown Mongolian metals market, global investors closely monitored the rush for gold and appeared enthralled by the vision that in Mongolia mining was like making "T-shirts for five bucks and selling them for $100" (Barta and Byambasuren 2007).[2] Vast amounts of money flowed into Mongolia, partly from exported minerals and partly from mining investments and financial speculation.

This explosive growth was far from limited to the formal mining sector and its many, predominantly foreign, companies. Thousands upon thousands of Mongolians also joined the hopeful search for the precious "yellow stuff" (*shar yum*) in the country's gold rush (see figure 1). Relying on cheap and basic equipment, ninja mining did not require lengthy training or substantial financial investment. The work routine was often impromptu and extemporized, thus not requiring a fixed team of miners working for fixed periods of time. Consequently, the gold rush constantly changed form and evaded any easy quantification and regulation. It rarely dwelled in any location for very long or comprised a stable population. As gold was excavated and moved along channels of the illegal gold trade, money flowed through the frontiers of rural Mongolia.

During my fieldwork in the district of Uyanga (Uyanga *sum*), stretching over two and a half years between January 2005 and August 2011, people kept a watchful eye on any movement in the gold price according to the London Gold Fixing. Reported daily on television news broadcasts just before the weather report, the price of gold appeared a completely ordinary topic of conversation among locals. For many, the influx of money to the region was seen as directly affected by changes in international commodities markets, currency exchange rates, and domestic inflation. There was a keen interest in and curiosity about money that was far from exclusive to the more usual specialists of financial speculators, trading houses, and economic advisers in the capital city.

Given this increased flow of money, it is perhaps not surprising that many commentators and advisers described the gold rush as "driven by poverty and desperation" (Grayson 2007, 1; see also Bazuin 2003, 4). It was depicted as a survival phenomenon for the poorest, who were falling between the cracks in a society that was pursuing a pronounced free-market strategy for wealth creation in the aftermath of state socialism (Tomlinson 1998; Rossabi 2005, 43).[3] Approaching gold rush mining as a reactionary poverty phenomenon is common beyond Mongolia and often voiced in circles of international development (Adepoju 2003; Weber-Fahr et al. 2001; Yakovleva 2007), popular media (e.g., NPR's four-part series entitled *Mongolia Booms* [Langfitt 2012]; the *Guardian*'s video series entitled *Mining in Mongolia* [Branigan and Chung 2014]; see also Hardenberg 2008; Barthelemy 2013; Aldama 2016), and international charities (for example, CARE, CAFOD [Catholic Agency for Overseas Development], ActionAid). Some commentators even refer to mining camps as "islands of prosperity in a sea of poverty" (John Hollaway in Labonne 1996, 118). As the gold price reached record highs, some analysts debated the extent to which the Mongolian gold rush could perhaps more appropriately be labeled a "wealth-seeking" rather than a "poverty-driven" phenomenon (Grayson 2007, 5), engendering new aspirations beyond satisfying mere basic needs.[4] Access to money was certainly central to the participants

Figure 1. A view across the gold mining valley of Ölt, Uyanga

in the Mongolian gold rush. For many ninjas in Uyanga, mining and money did relate to needs and demands, but it was rarely clear whether gold money fulfilled needs or instilled new demands, nor did people necessarily distinguish between the two. The problem with the conventional view of gold rushes is that it turns all rushes into similar examples of miners as victims of macroeconomic processes. The universalizing rationale ends up disregarding what is significant about the mining boom in Mongolia, namely, that while gold money relates to needs and aspirations, the miners themselves consider it much more than that.

From Rarity to Rush?

Many of the Mongolians I have met over the years find the gold rush more of a blemish, if not an embarrassment, than the "viable solution" to "the adverse effects of economic restructuring," as it was described by advisers to the country's government (World Bank

2004, ii). The Mongolian expression *altny hiirhel,* which I translate here as gold rush, captures the taintedness of the phenomenon. The noun *hiirhel* describes an emotional state in which a person wants something so badly that he/she is not concerned whether the pursuit is morally right or wrong. Considered out of control (*davraah*), such a *hiirhüü* person moves beyond or exceeds (*hetrüüleh*) the moral orders of reality (*bodit baidal*) in order to continue the intense pursuit of something—in this case gold. When I talked to people about the gold rush, they emphasized that today's gold rush was unique and morally contentious. As I took taxis around Ulaanbaatar, the drivers often positioned the gold rush as the cataclysmic endpoint in Mongolia's moral degradation. "What will the future bring?" they asked me rhetorically. In local shops where I bought my usual supplies before returning to my field site, shopkeepers shook their heads and sighed when the conversation touched upon the topic of ninjas, invariably bringing forth some uncharacteristically strong emotions. "Ninjas aren't people. What they do is wrong [*buruu*]!," a woman once exclaimed. In Uyanga, where the local government administration was faced with an intense influx of unregistered migrants and mounting budget constraints, I tried to escape the government officials who always took an encounter as an opportunity to launch into lengthy diatribes about the impossible state of affairs. They complained that "it was so much easier before" because "people wouldn't dig for gold. It was seen as bad [*muu*]. Bad things would happen." One afternoon, as I was helping my host mother on the steppe prepare the staple meal of noodle soup with mutton, an elderly herder who was a close family friend wondered, as if to himself, why "the environment has become so strange. In the old days we used to have abundant trees and nuts here—just a lot more than now. But now, after this mining, it isn't growing anymore." During my fieldwork I encountered various other, much less common, narratives. A drunken man once insisted that ninjas were evidence of a particularly Mongolian Chinggis Khaan-like fighter character. Another contemplated that the freedom of ninjas to roam, following the gold from one valley to the next, was an example of the long-standing desire for a nomadic lifestyle among Mongolians. These nostalgic narratives are not trivial or inconsequential (Stewart 1988; Berdahl 1999, 2000). Instead they produce a heightened sense of mining's importance in the present, dramatizing what has happened and what could potentially happen in the future.

In order to make better sense of the proliferation of moral narratives sparked by the mining boom, it is pertinent to take a brief look back in time. Rather than attempting to fill in the gaps in the often emotional narratives, archival and other historical material indicates recurring tensions and conflicts as well as transformations and new beginnings that have been significant in Mongolia over the past few centuries. As will become clear, the widespread longing for a past without mining is not simply a present critique of, say, postsocialist instability, global financial volatility, or contemporary free-market euphoria. It is rather part of a much more long-standing historical declamation about moral exemplarity and a waning way of life that is today considered "truly Mongolian" (*jinhene Mongol*) (see also Humphrey 1992).

Many of the concerns that are raised about gold mining today also arose in the mid-nineteenth century (see High and Schlesinger 2010). Back then, the gold fever had gripped the world's imagination with huge discoveries in California (1848), Australia (1853), British Columbia (1858), Siberia (1860), Tierra del Fuego (1884), South Africa (1886), and Canada (1896)—to name but a few. As these gold rushes spread, the largely unchartered territory of Inner Asia presented international investors, entrepreneurs, and adventurers with new and exciting prospects for further gold discoveries. But the emperors of the Qing Empire (1644–1911), which was the last dynasty of China stretching across today's China, Mongolia, and Tibet, were as a matter of principle against opening mines anywhere in the country. The Qing government judged that the local population was unlikely to benefit from the mining, which was deemed to attract only migrant workers prone to rabble-rousing, licentiousness, and violence (Kuhn 1990; Sommer 2000). These social concerns were matched by an equal desire to protect gravesites from disruptive digging and preserve the general purity (*ariun*) of the landscape.[5] As a result, the court repeatedly rejected petitions to open new mines whenever a formal request was submitted.

The government-level denunciation of mining was in many cases supported by the local population. When the Scottish missionary James Gilmour traveled across Mongolia on foot in the 1870s, he found that people were suspicious and hostile toward him every time he dug a few stones of interest out of the ground. He realized that they feared he was searching for gold like so many other foreigners at the time. If gold was mined, it was said to "take away the luck of the land" (Gilmour 1883, 190). Living on land that was considered "enchanted," the local population regarded mining as "a curse to the land and the people" (ibid., 255). This hostility not only concerned the act of mining but also applied to other activities that involved digging into the ground. According to the British diplomat Charles W. Campbell, the local population was fearful of the noblemen's stone palaces and the foreigners' new houses that were cropping up in Ih Hüree (present-day Ulaanbaatar). The sturdy walls that rose tall and the deep foundations that sat permanently in the land lent the buildings an ominous presence. "Mongols believe that all these old palaces and cities are haunted, and that the spirits are prone to bring misfortune on intruders" (Campbell 1903, 494).[6] The imminent presence of spirits in the land rendered the activity of digging transgressive and perilous for human residents.

However, toward the end of the nineteenth century the Qing Empire was beset by violent rebellions, fiscal crises, and aggression from imperialist powers. Some of the empire's most eminent statesmen began to suggest that the strict mining policy had grown out of touch with the demographic and economic reality on the ground. The demand for natural resources, particularly precious metals, was booming, and the empire's statesmen drew on European, North American, and Russian models to promote industrial mining as indispensable to the reconstruction of the decimated empire.

The first industrial mine to open in present-day Mongolia was Mongolor, which opened in 1899 with start-up capital from Victor van Grot. He was an ambitious Russian officer in the customs house at Tianjin, the port city east of the capital, Beijing. He ordered mining machinery to be sent from Europe at great expense and began contracting with workers. His contract explicitly stated that Russians were to be excluded from the labor force in favor of Chinese and Mongols, so he tried to hire locally. But this turned out to be a major challenge. No one wanted to work in the mines, and Mongolor had to "import Chinese labor from 1,000 miles away, simply because the Mongols refused at any price to work in the mines" (Montagu 1956, 77).[7] In the words of Frans Larson (1930, 245–46), who worked with van Grot between 1900 and 1902, "There was plenty of gold in the locality. The Mongols looked on interestedly at the work, but did not cooperate in it. Digging great quantities of gold out of the earth did not appeal to the Mongol as a profitable way in which to spend his days. Laborers had to be imported. We got some Russians, but most of our workers were Chinese coolies who came from Shantung."[8]

As a result of political tensions between the Qing and Russian governments, the arrival of Russian migrants to the Mongolor mines strained the political goodwill for the mining venture that was meant to bring some much-needed capital to the state budget. The local population was upset, demanding the mines be closed, and the government eventually decided to obey public opinion. Even if the "pure" land was no longer an imperial category for environmental preservation and control, it remained an important category to large parts of the population. Positioned centrally between increasingly diverging interests, the purity of the land was a concern that, for many Mongols, extended far beyond the political and economic aspirations of the Qing court. As for van Grot, Russian archival sources reveal that he ended up ruined by debt and was ultimately forced to flee to America "without a penny to his name . . . as a common laborer" (Romanov [1928] 1952, 510). The many years of local protests and negotiation had drained the investor's finances, and his bankruptcy served as a cautionary tale against gold mining ventures among the Mongols.

Decades of scant interest in Mongolia's mineral riches followed, and a systematic nationwide mapping of the country's geology was not carried out until 1960—four decades after Mongolia had strengthened and formalized economic, military, and political ties with the USSR. In the 1960s gold deposits were documented in all regions, detailing more than three hundred specific locations of major finds (Janzen 2005, 12). In 1962 Mongolia became a member of the centrally planned socialist trade organization COMECON (Council for Mutual Economic Assistance), and foreign investors arrived in the country with imported laborers who could work alongside the slowly growing Mongolian workforce. In this period, the singular industrial showpiece was the largest copper and molybdenum mine in Asia: the Erdenetiin-ovoo deposit.

The deposit was discovered by Mongolian and Czechoslovakian geologists in the mid-1960s and was developed in the following decade with massive financial and technical assistance from the USSR. Erdenet's development required the construction of a railway line, a highway, a water pipeline, and an electric power line connecting the area directly to the USSR (Worden and Savada 1991, 141). A Mongolian-Soviet construction force of fourteen thousand workers built the mine, which went into operation in 1978 with a planned annual capacity of twenty million tons of copper concentrate (Trifonov and Krouchkin 2000, 46). Housing facilities, schools, clubs, and medical centers were built, and by 1982 the town of Erdenet had a total population of forty thousand inhabitants with an average age of just twenty-four years (Nordby 1987, 120). Erdenet was *the* showpiece of socialist industrial life: the mine exceeded production targets and epitomized Leninist visions of heavy industry (Lkhamsuren 1982, 470; Sanders 1982). Many people moved to the town and found jobs alongside qualified Russians, who numbered one-quarter of the town's population. By the mid-1980s, mining was one of the most important economic sectors for the country and reportedly accounted for 42 percent of exports in 1985 (Worden and Savada 1991, 141), the main recipient being the USSR.[9] However, as Mongolians today like to emphasize, there were apparently very few informal-sector miners at the time—if any at all.

In 1990 Mongolia severed its ties with the disintegrating USSR, and the country embarked on a rapid program of privatization and democratization (Rossabi 2005, 45–62). During the first decade of "shock therapy," the country's crashing economy relied heavily on two economic sectors: pastoralism and mining. As people lost their jobs in the few urban centers, many city dwellers moved to the countryside in search of a new subsistence. The extreme continental climate and rural way of life posed a challenge to many newcomers from urban areas. Yet nomadic pastoralism was heralded as the savior of the Mongolian economy in the tough transition years. Thousands of people were absorbed into the large herding sector, assisting with the production of the main cash crop: cashmere. At the same time, the government began to make it easier to explore the country's mineral wealth that had been so well documented by the Russians. Mongolia's first minerals law was issued in 1994 and focused exclusively on formal-sector mining. Addressing licensing, exploration, and mining, the law provided a highly investor-friendly framework that laid the foundation for the impending explosive growth in mining (see High 2012). Between 1991 and 2001 global gold production grew by 21 percent. During the very same period, the gold production in Mongolia grew seventeenfold (GFMS 2004, 1). But even as the formal gold mining sector surged and led to dramatic increases in production, it was still relying on inefficient mechanized mining and processing systems. The technical equipment was originally developed in the early 1930s in the USSR and was not only outdated but also intended for mineralogical conditions that were unlike those in Mongolia. Consequently, much gold was effectively lost by the Mongolian and Russian-owned mining companies in these years. An authoritative report on Mongolian mining described the situation as follows: "The mining methods and gold washing technology of the mining companies has generally remained poor, and it is unusual for a wash plant to recover >60% of the gold. Once allowance is made for other gold 'left behind' (e.g. in side-walls, mine floor, overburden, etc.) then >50% of gold is estimated to have been 'left behind'" (MBDA 2003, 37; see also Samykina et al. 2005).

Using such poor technology, the formal mining sector effectively provided rich and relatively easy accessible resources for informal-sector miners at the same rate as it generated its own gold output. As the formal mining sector expanded in terms of investment, geographical spread, and gold production, the informal mining sector therefore grew accordingly. When droughts and disastrous weather conditions known in Mongolian as *zud* occurred between 1999 and 2002 (see chapter 1), the open placer mines (shallow alluvial deposits) were a logical destination, able to accommodate a new and growing wave of economic migrants.

In the early years of the gold rush, the International Labor Organization (2004, 1) released a report stating that it estimated around one hundred thousand people were already involved in the growing gold rush. In the years following this release, the number grew considerably. As the mined gold was sold along illegal channels, circumventing the 5 percent royalty tax on gold in place at the time, the Bank of Mongolia was losing much potential revenue from informal-sector mining. Other major income-generating minerals included coal, copper, fluorspar, molybdenum, tin, uranium, and zinc. In a legal climate that remained favorable to investment, mining had for years been the fastest-growing economic sector in the country, and at a stage of maturity still increasing its output by more than 33 percent per year (National Statistical Office of Mongolia 2006, 203).

In a state of exhilaration, an international investor once told me that he thought the Mongolian mining legislation was one of the most open and favorable in the world. In his view, it contained very little "restrictive legislation" in terms of government involvement, tax structure, environmental rehabilitation, and public arbitration. Indeed, the World Bank (2006, 2, 26) considered the regulatory and institutional framework for managing mining so "simple" and "weak" as to jeopardize the industry's long-term potential in the country. Given the magnitude of the formal and informal mining sectors, combined with a strong international interest in the country's mineral wealth, street protests, televised debates, and nongovernmental organizations often voiced their opposition to the government's keen stance on mining. Whereas some Mongolians supported the growing internationalization of the mining sector, others accused corrupt politicians of selling Mongolia's mineral resources and destroying the cherished countryside without generating significant public revenue.

Although mining has been a contentious topic in Mongolia for centuries, many people today draw on nostalgic narratives to position the current gold rush as historically unprecedented and unique. In a present context in which many feel that "people have completely lost their love for the landscape," as a herder in Uyanga commented, these narratives place moral traditions (*yos zanshil*, lit. custom and habit) in opposition to contemporary debates and controversial political agendas. At a time when the country is becoming heavily dependent on its natural resource wealth, the object of these narratives is not the recovery of a past with all its recurring tales of prospecting and protesting. It is rather the recuperation of a past without mining, a time when the land was once pure (*ariun*) and life was wonderful (*saihan*).

"No One Here Is a Miner"

Over the years my visits to Mongolia have always ended up gravitating around mining. I first went to Mongolia in the year 2001 as an intern for the International Labor Organization. My job then was to monitor development projects on child labor in the poorest parts of Ulaanbaatar, and it was not long before I was spending time with children scavenging for coal in abandoned coal mines on the outskirts of the city. From succeeding visits to Mongolia—first as a student at the National University of Mongolia and then as a doctoral student of anthropology—I witnessed the unfolding of the gold rush. In preparation for my first period of fieldwork in the gold mines, I spent a couple of months working as a volunteer for a Mongolian advocacy group called the Ongi River Movement (Ongiin Golynhan). It was established to generate awareness of the environmental impacts of mining along the river Ongi (see chapter 2). As my friendships with the members of the movement deepened, they invited me along on their field trips. Traveling from the South Gobi province (Ömnö Gov' *aimag*) to the Hangai Mountains, we surveyed mining impacts over several hundred kilometers and interviewed countless people about the pressing issues that mining raised for them. When we came upon the astounding concentration of gold rush miners at the head of the Ongi River in the district of Uyanga, I knew that this was to be the place for my fieldwork.

When I embarked on fieldwork in this so-called ninja capital (*ninja naryn nislel*) in 2005, I initially wanted to investigate the political tensions that I thought would emerge between local herders and the thousands of incoming gold rush miners. I had imagined a clear and explicit bifurcation between the two populations: on the one hand, incoming migrants who honeycombed the land for gold and turned rivers into stagnant mud from their mining activities, and on the other, local herders who relied on that very land and water for their own and their animals' livelihood. But when I arrived, I soon discovered that it was a very different reality.

I started fieldwork with a relatively wealthy herding family. Like my second host family on the steppe, it was an emphatically patriarchal household, of which the father, here known as Nyambuu, was the figurehead (cf. Sneath 2000; Vreeland [1954] 1962) (see figure 2). He made all major decisions, and his decision-making power extended beyond their children and grandchildren to incorporate all affines who lived in the household (as well as the visiting anthropologist). Their herd of yaks, horses, sheep, and goats numbered several hundred, and their *ger* (a round felt tent, known in English by the Russian loanword *yurt*) was lavishly decorated with beautifully painted furniture, shelves stacked with goods, and walls adorned with large carpets depicting various versions of a stoic Chinggis Khaan. My new hosts were often hosting visitors, offering them salty milk tea (*süütei tsai*), fermented mare's milk (*airag*), and home-brewed vodka (*shimiin arhi*). On these visits, gossip was exchanged and laughter filled the air. When it was time for the visitors to leave, my host mother, Degidsüren, invariably opened the household chest and dug out some boxes of chocolate and some money. "Then our visitors will travel well," she explained. Considering just how many people visited them, I was struck by my hosts' generosity. Thinking about how much money they parted with on a daily basis, on top of their other expenses, I asked on one of the first days whether anyone in the family perhaps sometimes worked in the mines. Degidsüren quickly replied, "Oh no! No one here is a miner." I spent the next many weeks milking yaks, preparing dairy products, collecting firewood, picking berries, drawing water, and cooking meals. At first there was no indication of ninja mining other than visitors who arrived with calloused hands and muddy boots.

Figure 2. Uyanga household head

One morning, as we sat around the stove after finishing the milking round, Nyambuu suddenly announced that they would be visiting their oldest son. I asked if I could join them, and after a lengthy negotiation he reluctantly invited me along. We drove in a packed car through the valley, and I was surprised to see how close the mining area was to my host family. When at their place, you could see only the tall mountains, the expansive tree cover, and the depression in the landscape where the river cut its path. No ninjas, no dirt mounds, no industrial trucks. And yet only a few kilometers away, the first ninja compounds appeared. As we drove on, more and more ninja clusters emerged. People covered in dirt walked between the closely set gers, carrying panning bowls, rubber mats, and bags full of gravel. We continued along a single dirt track with mining holes on either side when the mountainous landscape suddenly opened up to a huge area packed with gers. We had arrived in the large gold mining area of Uyanga, locally known as the "land of dust" (*shoroony gazar*) (figure 3). After stopping a few times to ask for directions, we finally stopped in front of a ger with a small sign announcing that it was a shop (*delgüür*). "I will introduce you," Degidsüren said in a voice that revealed slight trepidation. As we entered the ger, formal greetings were exchanged, and the ensuing conversation remained limited and stunted. Within ten minutes we were back in the car, on our way home. Despite the brevity of the visit, I gathered that Nyambuu and Degidsüren's oldest son was a gold trader (*altny chanj*) and that some of the family's other children, as well as Nyambuu himself, worked periodically as ninjas. The mother also sold meat and dairy products in the mines. And her older brother was one of the few so-called big bosses (*tom darga*)—that is, an illegal gold trader positioned above all the local traders who sells the gold in the capital and beyond (see chapter 6). Despite the initial denial that anyone was a miner, the family turned out to be centrally involved in the gold rush.

When I later moved in with another host family on the steppe, headed by Yagaanövgön, and began to follow ties of kinship and friendship between different families in Uyanga, I encountered the same kind of concealment and secrecy regarding mining. Being involved in the gold rush was not something people willingly disclosed. Indeed, a favorite pastime in the area was debating whether or not someone worked as a ninja. Working as a ninja meant that they had access to gold money, which should ideally be converted as quickly as possible. It also meant that they were physically close to, or perhaps involved in, the activity of extraction, in which of the "power of gold" (*altny chadvar*) and the anger (*uur*) of spirits were regarded as particularly strong and dangerous. This exposure made ninjas magnets of the feared "misfortune of gold" (*altny gai*), which could travel across the mines and bring calamities onto anyone who crossed its path.

As I began my fieldwork in the mines, my hosts tried to protect me against these dangers by giving me protective necklaces to wear, burning juniper incense in the ger, and reciting Buddhist mantras to ward off threats. My hosts were aware of the imminent perils of the mines. Indeed, it was altny gai that was said to have caused a serious accident during my stay that permanently debilitated my ninja host mother Tsetsgee. It was also altny gai that was diagnosed to have caused my sudden bout of illness during the winter of 2006. Life in the mines was intense. There were accidents, violence, and heavy drinking among both men and women. There were landslides, tunnel cave-ins, and soil erosion that frequently cost human lives. As a result, safety issues strongly affected how I carried out my fieldwork. I never went to the mines unaccompanied,

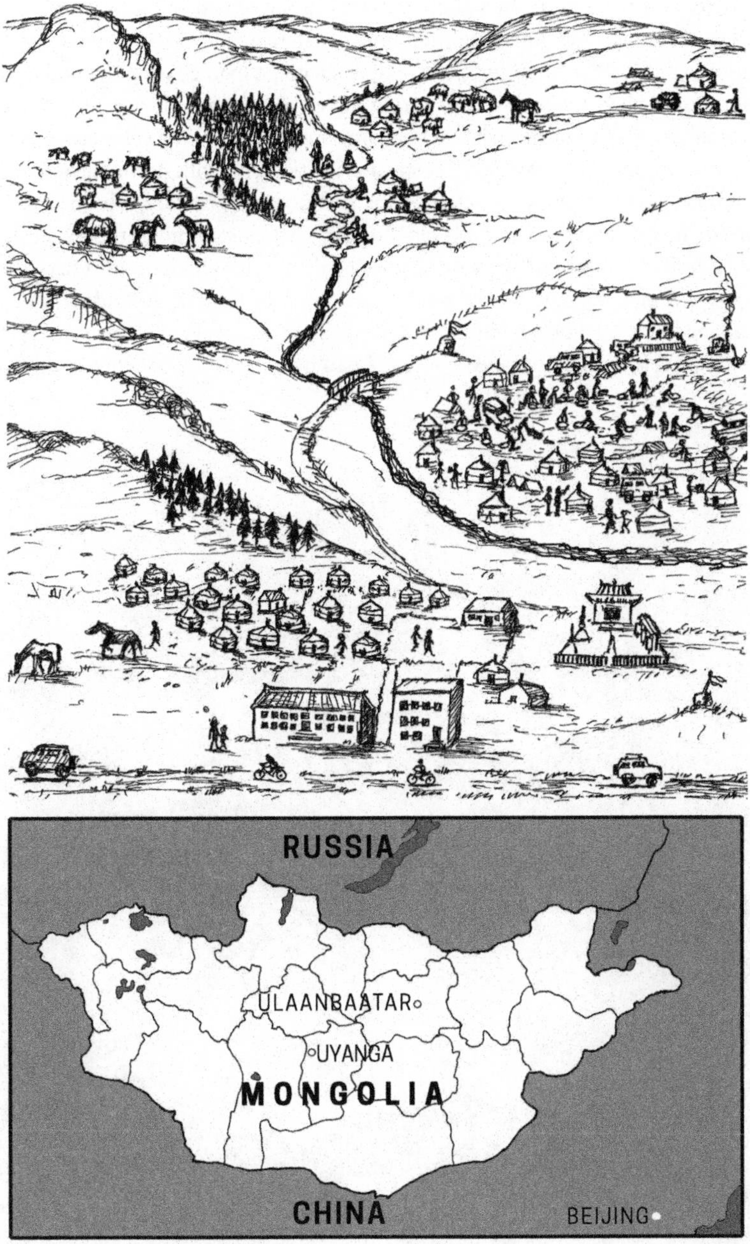

Figure 3. Map of Mongolia and Uyanga with the village, the mines, and surrounding herding area. Credit: Jos Sances, Alliance Graphics.

and because ninja mining was illegal, I had to ensure that people did not associate me with the police, the government, or any other formal institution. If they did, they would probably not have been willing to talk to me, and I suspect my stay in the mines would have come to an immediate end. Indeed, when I made my first, and in hindsight rather foolish, attempt to undertake a simple survey, a man chasing me with a raised fist put a quick halt to that kind of formal data collection. This meant that I could hardly make use of research assistants or carry out recorded interviews in the area.

Accepting such compromises, I instead prioritized my long-term participation in and observation of daily work-related jobs and domestic tasks. Both on the steppe and in the mines, I worked and relaxed alongside my host families, their friends, and relatives. In the mines these were male and female miners, children, "dirt taxi drivers," local shopkeepers, fortune-tellers, bonesetters, monks, illegal gold traders, and many others (see figure 4). I began to purposefully train my memory skills, especially focusing on memorizing sentences. As soon as I had had a great casual conversation with

Figure 4. The mining landscape of Uyanga

somebody, I would try to return to my host's ger and write it down. However, in the mines bags of gravel would quickly pile up if my pace at the sluice box or by the panning lake slowed down, and I sometimes had to wait for hours before I could leave to write notes. As my language skills slowly improved, I increasingly used Mongolian words and sentences in my notes. Not having to translate made memorization easier in some ways, but it also helped focus my attention on local expressions, idioms, and slang. Moreover, it sometimes gave rise to secondary data when I decided to share my notes with my hosts. Great conversations were sometimes sparked and deeper understandings could be sought.

In addition to fieldwork with herders and miners, I also stayed with local Buddhist monks (*lam*) in the nearby village (*sumyn töv*) of Uyanga. In 2006 the village itself had a population of around five thousand people and also housed Uyanga's monastery (*Uyangiin hiid*), the local school, a basic hospital, local government offices, and small shops. During my stay with the lamas, they generally allowed me to record conversations, and they shared with me many accounts of people's various concerns. In an area where anyone might be a ninja and gold money was constantly circulating, people's concerns were numerous and wide-ranging. Outside the monastery and the lamas' compounds, these delicate matters were carefully broached only among trusted interlocutors, if at all, and then in a subdued whisper in order to prevent spirits from eavesdropping. Any confrontational statement heard by spirits could potentially upset these beings and invite even greater calamities. Or, at a minimum, it would offend them since they ought to be treated just as respectfully as any other human or nonhuman being (see also Humphrey 1995, 137; Højer 2004). In a region where the land was lamented as having lost its purity for the sake of generating gold money, people were faced with a reality in which the gold rush was far from simply about satisfying the mere basic needs and demands identified by government advisers and development agencies. It was a radically emotive phenomenon by which people were confronted with a reality of both fear and fortune.

The Money Object

As anthropologists, one of the problems we encounter when studying money is its immediate familiarity. In places where state currencies circulate, minted coins and legally tendered bank notes can appear familiar and comforting to us in a setting where everything else might seem dizzyingly unusual and puzzling. Yet its familiar material form might belie a much more interesting and unexpected reality. At least that is the case in Mongolia's gold mines.

In 2006 amendments to the minerals law were under discussion and eventually approved. At the time there were intense public debates and protests in Ulaanbaatar. People organized hunger strikes, pitched gers in the central square, and burned effigies of the then president, the speaker of Parliament, and the mining tycoon Robert Friedland. Strong emotions were on display across the city, highlighting the animosity of many toward the mining industry and the support it received from the government.

Perhaps responding to the vocal public, if not also heeding the recommendations of some international advisers, the government decided to increase state control over its mineral wealth and introduced "compensation payments" (*nöhön tölbör*) to affected land users (article 41). In Uyanga, this institutionalization of monetary payments from mining companies to local residents received a rather mixed welcome. "We don't want money from a gold mining company. . . . That money has no good origin [*üüsver*]," Yagaanövgön warned. Another herder mumbled to himself that he "didn't understand well this money" (*ene möngö sain oilgohgüi*)[10] and asked me to what extent such payment was common across the world. Although hundreds of households in Uyanga were directly affected by mining, the local government office told me that only three had taken the compensation payments. "People here don't want this money," the governor said. As a material object, the Mongolian state currency is earmarked in ways that affect its desirability. The linkage between money and gold mining is not an abstract matter of moral condemnation but part of everyday economic life. It is a differentiation that affects the liquidity of money, that is, the ease with which it can be used in transactions. Although classical monetary economics position cash as the most liquid asset of all because it can be sold instantly without loss of value, Uyanga's gold money has become a controversial object of disjuncture—an object that is vulnerable to sudden and unpredictable shifts in value.

The anthropological treatment of money has, since its early days, devoted much attention to the search for cross-culturally applicable definitions—those that also apply in the absence of a singular currency or nation-state. Objects such as salt bars, shells, and beads have been shown to operate in ways that were at once similar to, yet distinct from, Western currencies. Instead of focusing on the immediate material differences of the objects used in exchange (Graeber 1996), early exchange theorists such as Malinowski (1922), Mauss (1925), and Firth (1929) brought attention to the underlying shared logics that inform economic relations. They sought to highlight the similarities in their function and provided detailed typologies of various forms of reciprocal acts (see Sahlins 1965). This interest in "objects as currency" offered potentially important revisions to Western monetary theory (e.g., Hogendorn and Johnson 2003; Keynes 1982). It also elucidated and elaborated the multiple functions of money as defined by Aristotle and still often repeated today in anthropology and economics textbooks: a store of wealth, a means of payment, a unit of account, and a medium of exchange.

This focus on the function of money in anthropology has contributed to its positioning within a dyadic framework that highlights the agency of its holder: people handle money in order to achieve certain ends, whether material or immaterial. Positioned as a facilitator of human desires, money is often characterized more by what it does for its holders than by what kind of object they consider it to be. As the analytical attention is directed toward the functions of money and their broader implications for social relations, we risk overlooking how currency actually becomes exchangeable. Just like any other object, currency circulates through various regimes of value (Appadurai 1986, 4; Kopytoff 1986); sometimes it is a powerful symbol of gratitude and respect in a hospitality ritual, while at others it is considered first and foremost a token of cash value (Humphrey 2000, 234–46, 2012; Sneath 2006). As a material object, currency is subject to interpretation and contestation, and there is always the possibility that it might slip into alternative regimes of value. Or, as Keane (2001, 69) puts it, "The status of money itself is not entirely stable." As a social medium, money is subject to multiple and conflictive interpretations that centrally compete with its function as currency.

When anthropologists have dealt with money as an object, they have demonstrated its cross-cultural symbolic capacity in moral evaluations, giving rise to its status as "the symbol of all symbols" (Gregory 1997, 33). Bloch and Parry (1989, 3) have argued that the tendency to construct money symbolically as "an incredibly powerful agent of profound social and cultural transformations" is deeply rooted in our *own* cultural tradition and is reproduced repeatedly in Western social thought.[11] Even today, as Maurer (2006, 17) notes, we have remained loyal to the promises offered by the evocative symbols of devils and demons, narratives of great transformations, and the teleologies of decline and despair. Money's potential as a symbol to encapsulate processes of transformation lies at the very heart of the so-called occult economies in which people deploy "magic currencies" in pursuit of otherwise unattainable ends (Taussig 1980; Comaroff and Comaroff 1999; Geshiere 1997). In people's attempts to navigate "'millennial capitalism' with its fusion of hope and hopelessness, promise and its perversions" (Comaroff and Comaroff 1999, 283), money becomes both an index and a source of powerful occult agency.

But when we recognize it as an object typified by exchange specifically and look beyond its striking capacity for symbolic representation, we see that money is further entangled in the uncertainties of interaction. As Strathern (1988, 1992) has shown in the context of Melanesian gift exchange, we have long been seduced by the magic of the "barter model of value": a model in which objects exist independently of an exchange and are transacted in response to people's needs (e.g., Simmel [1907] 2004). Each party is defined as supplying what the other wants, and as a result there is no problem in establishing the substitutability of the exchanged items. The intrinsic, separate qualities of things and persons ensure that the challenge in transactions is simply one of establishing an exchange ratio.[12] However, the presumption that "the quantity of one thing can be enumerated in respect of that of another is the magic of reification" (Strathern 1992, 172; see also Guyer 2004; Fourcade 2011). And gold can be seen as the most magical of all such reifications: it is presumed exchangeable with all other objects and presumed a potential universal store of value. This is why economists turn to gold as a recommended asset for investment in times of fiscal crisis (but see Feldstein 2009). Whereas such reification might make sense within some Western, especially Friedman-inspired, ideologies, how can we assume that people universally share such views? If we approach objects, including currency, as not a priori separate and bounded, questions of money, gold, and cash value become less straightforward. In fact, if currency is not a stable instrument of equivalence, what is the value of a transaction? And when gold is considered in the same light, how can one talk about the cross-cultural value of gold? To what extent does the reality in mining camps relate in financial ways to the global interest in gold? These issues have arisen in the Mongolian gold rush, in which the seemingly uniform national currency of *tögrög* is locally regarded as not all the same, precisely because it does not lend itself well to the magic of reification. The macroeconomic ideal of money, including the physical currency, as a transparent, stable, and singular tool in exchange is a vision that is not always shared by all its holders (see also Zelizer 1997).[13]

Although quantity theorists in monetary economics approach the reified abstraction of currency as a neutral, if not benign, construction, it is crucial to remember that it is dressed in a "national uniform" (Marx [1987] 1967, 125) and carries emblems of the state (Hart 1986, 2007). As Gregory (1997) emphasizes in his study of colonial currencies, money was historically created by kings as *their* standard of value. It was implemented as a means of fixing the prices of special commodities such as precious metals, beads, and cowrie shells, and with the intention of facilitating the levying of taxes (see also Guyer 1995; Mintz 2007). If people regard currency as an object that is not divorced from the social world, the fixing of prices and the extraction of taxes can be seen as a profound imposition of political power. And this imposition is particularly intense in practices of extraction, circulation, and use of minerals. As noted by Ferry (2005, 6; see also Godoy 1985, 207), studies of mining have often documented the centrality of the state in the strategic organization of mining operations (e.g., Golub 2014; Rajak 2011; Saunders 2014) as well as in the daily life among the miners themselves (e.g., Donham 2011; Kirsch 2007; Nash 2001; Robinson 1986). However, in the Mongolian gold mines state powers seem distant, if not irrelevant, to how life is lived. State presence through taxation and regulation is highly intermittent, and state actors appear virtually absent. Here currency is associated not only with the state but also with multiple local powers, circulating as a potent object that highlights the intense mutual implication of people with other human and nonhuman forces.

Anthropology has had much to say about gift exchange and reciprocity over the years, yet we have had surprisingly little to say about the fiscal capacity of national currencies (though see Foster 1998; Guyer 2004; Maurer 2008). Through extensive revisionism we have learned that establishing equivalence between different objects is hardly a straightforward and simple task (Beidelman 1989; Foster 1990; Myers 2001). I wonder why it should be any different with national currencies (see also Keane 2008). This book calls for critical attention to the ways in which the often important reality of money makes sense to people locally. Alongside official doctrines and Western monetary economics, people's own understandings form a key part of how the money object works in practice.

Cosmoeconomies of Extraction

Recognizing that currency and other wealth objects are malleable media, how are we to approach this malleability analytically? As the Mongolian tögrög intersects with spirit beings that move through the pockmarked mining landscape, how can we understand the relationship between spirit worlds and human lives? The forceful imposition of human agency in extractive economies often spurs a host of contentious issues. Suddenly, as in so many other parts of the world, large corporations, national representatives, and local stakeholders are brought into tense proximity through their shared, but rarely similar, interests in a development. Some of these encounters become "constructive dialogues" (Strang 2004, 219), others turn into "resource wars" (Ballard and Banks 2003, 289) involving major court cases over compensation payments, royalty payouts, and rights abuses. But these social dramas are not necessarily premised on the autonomy of human agents. Multiple kinds of beings sometimes have a vested interest in the same oil (Stewart and Strathern 2002), lay claims to the same minerals (Sallnow 1989, 212), or inhabit the same forests designated for logging (Wood 2004, 251–52). As extractive industries physically overturn the landscape in their search for natural resources, they create financial flows but also, as in the Mongolian case, related cosmological transformations.

This imposition of human agency is a characteristic not only of extractive industries but indeed of our very moment in geohistory. The Nobel prize–winning atmospheric scientist Paul Crutzen (2002) and his team of geoscientists have shown how human activities are now exerting such transformational impacts on the atmosphere that they in many ways outcompete natural and other processes. As Bruno Latour (2013) noted in the Gifford Lectures at the University of Edinburgh, no postmodern philosopher, no reflexive anthropologist, no liberal theologian, no political thinker would ever have dared to weigh the influence of humans on the *same* historical and analytical scale as that used for rivers, floods, and erosions. But for the geoscientists, our world has now entered a new and undeniable era: the Anthropocene—an era in which the collective anthropos has become an earth-changing force. Although we might contest its epochal diagnosis, the stratigraphic term is significant because it poses a radical and fundamental challenge to the modernist understanding of nature as a pure, singular, and passive domain distinct from a human-agentive society (Latour 2009, 2014; Whitehead 1920). More than offering a dialectical reconciliation between nature and society, the notion of the Anthropocene can be seen to completely bypass the separation. Neither nature nor society can exist whole and intact in this new epoch—that is, if the separation, supposedly terminated by the Anthropocene, ever actually *did* exist (see also Castree 2014; Lorimer 2012). In the Anthropocene, humans are positioned on center stage, reminding us that the "posthuman turn" does not imply any diminished significance of the anthropos. Instead, it signals the significance of all the various agencies that together make up the "pluriverse," to use William James's expression (James [1909] 1996), in which humans reside and act. Recognizing the proliferation of human and nonhuman agencies, it admits the possibility not only that we have "never been modern" (Latour 1993) but also that "we have never been human" (Haraway 2008; see also Tsing 2015).

Giving attention to posthuman possibilities of existence, in this book I ask how we can approach cosmologies in a way that recognizes more adequately the multiple and dynamic relational ties that exist within the Mongolian gold economy. To what extent can we make sense of spirit worlds and gold money without reducing cosmology to economics or vice versa? And what if we did not have to take recourse to such isolating and artificial categories in the first place (cf. Moore 2006, 23)? Indeed, people in Uyanga often do not make such distinctions themselves when describing their predicaments. Like many others in the Mongolian cultural region, their multiple and often mutually incongruent ideas about the human condition amount to much more than a discernible set of definitive propositions (see also Buyandelger 2013, 91; Højer 2009; Humphrey and Onon 1996, 76; Pedersen 2011, 180–82).[14] People rarely, if at all, tell all-encompassing cosmogonic myths that lay bare the origin of all things. Nor do they generally discuss their broader ideas about what the world is like in any direct or systematic fashion. Yet, as Michael Scott (2007, 5) notes, the absence of comprehensive myths and discernible cosmogonic narratives does not mean that their cosmology is thereby "inaccessible or unintelligible" (see also Mimica 2010, 212; Abramson and Holbraad 2012). Even if anthropology has traditionally looked to myths and ritual practices for keys to local cosmologies (e.g., Barth 1975; Hugh-Jones 1979; Viveiros de Castro 1998), these are not necessarily the only, or even principal, domains of everyday life where cosmologies figure. In Uyanga, it is specifically the extraction and transaction of gold, or in Michael Taussig's (1980, xii) words, "the human significance of the economy," that repeatedly gives rise to cosmological evocations.

In using the term "cosmological evocation," my approach builds on Christopher Healey's (1985, 1988) work on the cosmology of exchange among the Maring of Papua New Guinea. Seeking to understand the metaphysical and agentive continuities between spirits, animals, and humans, he looks at the fluid and "penumbral" associations of spirits with human society in everyday practices as varied as hunting, commensality, war sorcery, and gift giving. Emphasizing how cosmological evocations emerge contextually and partially, he observes that "the conceptual dimensions and content of cosmologies retain a somewhat inchoate character" (1988, 107). In contrast to the calculated and goal-driven logics presented in "spiritual economies" (Rudnyckyj 2010) and "occult economies" (Comaroff and Comaroff 1999), Healey thus brings to his analysis a sensibility that recognizes individual human beings and cosmological understandings as fundamentally dynamic and disparate. Rather than attributing facticity, and indeed abstract holism, to cosmological evocations, I seek to approach the interface of cosmology and economic life as an active challenge for my various hosts. As people in Uyanga were often keen to point out, a vast number of human and nonhuman beings are involved in the Mongolian gold rush, and this involvement is not one of consensus or harmony (see also De la Cadena 2010). It is also not an involvement that is entirely known or even knowable. As a result, it is important to avoid presuming a cosmos that is grounded in epistemological coherence and certainty. It is also important to avoid the notion of a cosmos that is a passive, ahistorical, and apolitical backdrop to our actions, emerging only periodically in myths and rituals. Rather, it is a cosmos that is centrally implicated in the tense symbiosis between humans and nonhumans as they now take part in Mongolia's contentious mining boom.

Fear and Fortune

In tracing the continuities and disjunctions between human and nonhuman participants in the gold rush, this book seeks to follow the paths of gold from the point of extraction to some of its many destinations. Structured to reflect this expansive cosmoeconomy, the chapters introduce humans as variously positioned as herders living on the margins of the mines, ninjas extracting the "yellow stuff" (*shar yum*), shopkeepers handling gold money, Buddhist lamas trying to mitigate impending misfortune, and finally illegal gold traders journeying far beyond the remote mining camps of Uyanga. Shifting across these different terrains, I explore how people maneuver within a financially profitable but morally charged economy of mineral extraction.

The following two chapters begin by examining people's understandings of the gold rush as an emotive phenomenon to either take part in or dissociate from. Chapter 1 outlines how disastrous weather conditions hit Mongolia between 1999 and 2002, throwing thousands of herders into sudden poverty and leading many of them to the mines. Yet many of Uyanga's herders who managed to sustain their large herds were indeed also attracted to the mining camps. Focusing on these wealthy herding households, I show how patriarchal relations affected people's everyday lives. Attending to frustrations and discontent within households, I show how "following gold" (*alt dagah*) is seen to offer a viable source of income that is relatively independent of the hierarchical power dynamics in Uyanga's herding economy. In this light, the gold rush is a pressing reality that offers an alternative way of life that appeals to many.

Since washing dirt relies on the central use of water, a major issue confronting herders is water scarcity and contamination. Introducing this troubled relationship between emerging and established economies, chapter 2 describes how the involvement in mining is seen as a fundamentally different and incompatible way of life rather than a complementary option to nomadic pastoralism. I examine how the invisible dust known as the power of gold, which emerges during the extraction of gold, draws people, sometimes against their own will, to the mining camps. Ever since the first mining company broke the soil and initiated large-scale gold mining, the gold rush has evolved into an unstoppable force that even national activists feel unable to bring to a halt.

Chapter 3 explores how people involved in gold mining rely on more than just knowledge of local geology and mining technologies in their search for gold. They also enter into relationships with new and dangerous spirit beings in order to attract and harness fortune (*hishig*). But as they do this, ninjas are said to also come into contact with the misfortune of gold, which can be redirected onto others through ritual practices. With illnesses readily blamed on ninjas, they are subject to much suspicion and enmity. Examining local understandings of a transforming landscape, I show how the desire for subterranean wealth demands recognition of powerful spirit worlds.

In chapter 4 I examine local understandings of wealth, specifically the relationship between pastoral wealth and gold money. I look at how the dangers surrounding mining are transferred onto the money object itself upon its sale to resident gold traders.

As a vector of pollution, money circulates as a material objectification of potential calamity. When using dirty money notes associated with the mines, people face a local redenomination and de facto lower purchasing power when spending their money. Although the gold rush has enriched ninjas, the cash value of their money is intimately tied to its materiality in local moral understandings of value. This chapter shows how the booming mining economy has given rise to fears of circulating misfortune, irrespective of people's own ties to the mines.

Chapter 5 considers how local Buddhist lamas have become involved in the gold rush. Within Mahayana Buddhism, mining is considered emblematic of violent acts and self-interest that infringe upon the desired relationship between people and the environment. Ninjas thus often call upon monastic religious specialists to carry out appeasement ceremonies and other rituals in the gold mines and in the village monastery. With most of their ritual services now directly related to the extraction of minerals, the monastery receives large amounts of gold money. This wealth is much welcomed as it helps sustain the village monastery to an extent that the monastery has never experienced before. Yet as junior lamas struggle to afford their monastic service, they have started to raise the issue of money's distribution within the monastery. In this chapter, I thus show how the lamas have become central to not only the local mitigation of impending misfortune but also the circulation of gold money.

Chapter 6 turns to the gold traders who take part in economic circuits that are oriented away from the mines and toward the yuan of their Chinese trading partners within Asia's illegal gold trade. For them the intersection of gold wealth with international money flows is conducive to the transformation of their "lifeless" earnings into profitable and productive currency. Holding and handling unmatched quantities, they quickly reinvest their "renewed" money into the gold trade or a business venture. Often transformed into visible, material wealth, money from the illegal gold trade thus offers a competing topography of wealth that is based not on the accumulation of fortune in livestock but on risk taking and business acumen. At a time of growing demand for gold in neighboring China, this chapter shows how different understandings of money's value and materiality intersect, supporting the liquidity of gold money and the emergence of a new hierarchy of powerful men in Uyanga.

Ultimately, the emergence of the cosmoeconomy described in this book demonstrates how a new way of life that gravitates around gold has been accompanied by an intensification in spirit presence, impending pollution, and ritual intervention. In contrast to accounts that consider religious resurgence to be either a refuge from or a means of resistance to economic globalization (Castells 2011; Comaroff and Comaroff 2000), Uyanga's various inhabitants show us a certain coexistence of spirit worlds and natural resource extraction. However, this is not a triumphant tale of their benevolent reconciliation. Rather, it is an account of the conditions under which global gold reserves are produced. At a time when the gold price continues to scale new peaks, the intense Mongolian cosmoeconomy reminds us that investment strategies to buy gold as a hedge or harbor against financial crises set in motion a profoundly uncertain future for many—both human and nonhuman.

1 The Burden of Patriarchy

"HAS YOUR DAD gone senile [*zönög*] yet?" Ahaa, the oldest son of Yagaanövgön,[1] asked as we drove across the desolate steppe on his motorbike. We were on our way to Ölt, also known as the land of dust (see figure 4 in the introduction). Up until a few years ago, Ahaa had lived and worked in the mines with his wife, but the hard work had eventually taken its toll on his back, and they had returned to his father's *ail* (household cluster). They now lived in their own *ger* with their four-year-old son, but living with Ahaa's family was not easy for them. "Our dad is now in his sixties and he's going senile, right? He's making really bad decisions [*shiidver*] but refuses [*tatgalzah*] to listen to anyone. He has always been like that, but now that he is getting older it is really hard to accept his decisions. But we have to! We have to accept all his decisions on everything and can't do anything on our own. I'm so sick of it [*zalhah*]!"

During my fieldwork in Uyanga, such voiced frustrations occurred quite often in the large herding households located in the vicinity of the mines, especially when daily tasks took us far away from the compounds. Ahaa's situation was particularly difficult. His father was one of the wealthiest and oldest herders in the area and immediate respect was expected to accompany his movements within and beyond the ail. While he was respected for being a nice and generous man, he was also single-handedly in charge of all final matters relating to the ail and its herd. Indeed, Yagaanövgön was a strong-minded man who enjoyed a good discussion that preferably confirmed that he was right. Although the village and the mines were located only thirty minutes away by motorbike or an hour by horse, they seemed very far away. The family's nearest neighbor lived on the other side of the mountains, and they sometimes went many days without seeing any visitors (see figure 5). In particular the eight months-long winter entailed a remarkable degree of isolation in which they drew their daily entertainment almost exclusively from one another's company. At such times, the elderly man could be seen walking alone behind his herd, bow-legged, leaning slightly forward into the wind and with his hands clasped behind his back. Or he sat in the honored northern part of the ger, lighting a "butter candle" (*zul*) on the altar before launching into a lengthy monologue about matters he considered urgent or simply interesting. At times he talked about the kind of medical treatment he thought necessary for his wife, who was suffering unbearable kidney pain. At others, he described the kind of tea he wanted his sons to buy upon their next visit to the village.

Figure 5. Living remotely

As he spoke, family members occasionally mumbled their discontent with his decisions. Given his poor hearing, it was likely that he did not notice their objections. Or perhaps he simply did not care. But when he did hear, he looked sternly at the person in question and exclaimed a forceful "What?!" (*yüü*). Regardless of the response, he mumbled exhaustedly, as if to himself, "Oh, how difficult this is! Even my own family doesn't listen to me! What to do . . . [*Üü, yamar hetsüü baina daa! Manaihan l nadad toohgüi! Yüü hiih ve* . . .]." More often than not, the other member eventually grew quiet, and the elderly man continued his speech. But as soon as an opportunity arose to momentarily leave the ail, household members would often vent their exasperation.

Before carrying out my fieldwork in Uyanga, I had presumed that ninja mining would come up most intensely in contexts that related to the obvious environmental clash between the country's two largest economic sectors: herding and mining. With thousands of ninjas panning for gold, rivers were turned into stagnant mud, leaving herders with no drinking water for themselves and their animals. As fertile pastureland was perforated with deep mining holes, connected underground with unsupported tunnels, the mines were unlikely to ever again become safe for human and animal habitation. These environmental disasters have indeed happened, and herders often lamented the seemingly gloomy future ahead of them. This reality struck during the winter of 2005–6 when herding households ran out of drinking water. Since my hosts lived upstream from the mines, their water was not polluted with heavy metals and sediments from panning, but waves of ninjas had begun prospecting at the headwaters of previously untouched streams that fed into the region. When the nearby spring one day dried out, people were entirely dependent on the intermittent snowfall. Bags were filled with snow and stored indoors until needed. Men equipped with axes and saws broke loose large chunks of ice that were transported back on horseback. When precipitation stopped in late February and there was no more snow or ice to collect, desperation took over and many households in the area began to move. In this situation my hosts furiously blamed the nearby mining for their feared predicament. However, in daily life such dramatic conflicts were rare, and complaints about the environmental consequences of mining were usually incorporated into much more casual conversations about the general difficulties of the herding life. Even if it was the environmental clash that struck me, these contexts did not spark the most frequent allusions to mining among people living on the steppe.

This chapter introduces the many herders who have been drawn to the mines. Apart from those who have lost their livestock in weather disasters (*zud*), I consider why members of wealthy herding households have also decided to follow gold (*alt dagah*). Rather than attributing the rush to a historical exodus away from nomadic pastoralism or poverty, the chapter shows how the search for gold is a pressing reality in daily life on the steppe, intersecting with and feeding off some people's desires for an alternative way of life.

A Gold Rush in Uyanga

According to official records, gold was first discovered in Uyanga in 1940, and in almost every subsequent decade further geological surveys were carried out by Soviet geologists (see Erel Kompani 1994). Davgaa and Zayat, an elderly couple, vividly remembered the geologists' visits: "The Russians came here a long time ago and took photographs. They brought all kinds of strange [*hachin*] equipment with them and walked up and down the valley. They stayed in their own tent and didn't come to visit us. But they kept coming back many, many times."

The Soviet geologists never built a mine in Uyanga, but they carried out all the preparatory work, which they filed in the GeoFund, the once-secret national archive for geological data in Ulaanbaatar. Decades later, when the archive was opened to the public, the Soviet reports were accessed by the Mongolian mining company Erel, which soon became the largest in the country from the fortune it made in Uyanga.

Whereas much of Mongolia's gold is found in hard-rock formations that require considerable time and effort for its extraction (Murray and Grayson 2003, 45–48,

Grayson 2006), Uyanga's gold is found in placer deposits.[2] These are "deposits which originated elsewhere and at a later stage ended up 'placed' in their locations, mainly by movement of water, but also by movement of wind and sand. Since they are relatively younger than their matrix, they are not geologically integrated with it and hence relatively easy to extract" (Stemmet 1996, 8).

This particular geological formation of gold requires only minimal technology for its extraction yet provides a relatively high yield. Given its accessibility, placer gold has been mined since ancient times and has been pivotal to many of the world's historical gold rushes. The simplest way to extract placer gold is by panning (see figure 6). A handful of mined ore is placed in a panning bowl and swirled in water so that the lighter material, known in Mongolian as *shalaam*, is washed over the side and the heavier gold flakes settle on the bottom. In Uyanga, panning is often used in conjunction with other mining techniques such as sluice boxes (*pünkyör*) and water cannons (*usan buu*). Sluice boxes are long, open-ended, often handmade metal cases. The bottom is lined with indented rubber mats (*erzeen*, also known as *ryezin*), and as the mined ore is shoveled into the slanting sluice box together with water, the ore washes through the box and the heavier particles lodge in the mats. Sluice boxes are often combined with drums (also known as trommels) or rockers (*pajur*) that, together with water, effectively sift the mined ore through holes in the side, allowing only the finer particles to enter the sluice boxes. All these mining techniques rely on the central use of water and, with the exception of panning, have all been used in Uyanga's large-scale hydraulic mining operations, which began immediately after the collapse of the socialist regime.

In 1990 Erel received government approval to begin mining only a few kilometers from the village in the area where the Soviet geologists had years before located large placer deposits (Erel Kompani 1994, 21). Mining exploration started in 1993 on the placer that runs through the valley adjacent to the Ongi River (see figure 1 in the introduction). At that time, the Ongi River was among the longest rivers in Mongolia, flowing from its source in the Hangai Mountains of Uyanga into the large lake of Ulaan Nuur in the South Gobi province. Diverting and setting up dams on tributaries to the Ongi River, Erel used the river water to feed its high-pressure water jets. The water jets were directed at the gold-bearing deposits, and the resulting watery sediment slurry flowed through large sluice boxes. Once the slurry had been processed, tall heaps of waste tailings were left behind. As with any hydraulic mining, Erel in effect washed out the entire valley and its hillsides to get to the underlying deposits, sweeping tons of debris into nearby streams and rivers.[3] Over time the debris settled, contributing to today's heavy mineral and sedimentation pollution of the Ongi River (Tungalag, Tsolmon, and Bayartungalag 2008; see also Lovgren 2008). Now nicknamed the Red River (Ulaan Gol) because of its changed color, the river runs for less than one hundred kilometers before drying up (see chapter 2 for a discussion of attempts to protect the Ongi River). As the mining company slowly worked its way up the valley, ninjas soon moved in and started panning the waste tailings for leftover gold.[4]

The Mongolian gold rush appears to have started around 1995–96 in several different areas, one of which was Uyanga (MBDA 2003, 25). That these areas are far apart likely

Figure 6. Children panning for gold. Credit: Jos Sances, Alliance Graphics.

reflects people's general desperation and creativity at a time of severe societal turmoil. State-owned mining companies had closed, and many miners suddenly found themselves without jobs. The GeoFund archive had just opened its doors to the public, and, as geologist Robin Grayson (2007, 3) notes, "The Soviet GeoFund enabled the placer gold rush to be a fast, precise, profitable mining rush; not a slow, speculative exploration rush, for the Soviets had proved a vast resource of ready-to-mine placer gold in more than two decades of churn drilling, bucket drilling and pitting."

In these turbulent transition years, many unemployed miners had the experience and technical knowledge to transfer hydraulic mining methods to small-scale, low-investment, artisanal mining operations. They also had ready access to shelves upon shelves of detailed mineral investigations, detailing the exact location and expected size of profitable deposits. As a report outlines, "Faced with a very uncertain future, a few thousand mining specialists chose to become informal gold miners, doing what seemed natural to them" (MBDA 2003, 25). This first wave of gold rush miners was then followed by people with diverse backgrounds from all parts of the country. In the words of Dorj, an elderly villager in Uyanga, "Suddenly people started coming. They came from all sorts of places, like Gobi-Altai, Bayanhongor, Hentii, Dornod, Arhangai, and elsewhere.[5] One hundred people came, maybe two hundred, or three hundred. They came with their gers and everything. We were invaded by outsiders [*gadny ulsuud*]."

A young female shopkeeper with whom I worked closely described the early gold rush years as a time when "almost the whole nation seemed to gather around here to dig holes." In Uyanga herders constituted the largest source of ninjas and many of them were from the local region of Övörhangai (see also MBDA 2003, 26). Around 1999–2000, when 36 percent of the population was officially recorded as living below the poverty line of 17 USD per person per month (National Statistics Office and United Nations Development Programme 1999), ninja mining in Uyanga was locally described as an *altny hiirhel* (gold rush). By then, thousands of people were mining for gold in the same valley as Erel.[6]

As the amount of gold in the waste tailings of the lower valley began to decrease, ninjas slowly followed the direction of the company's operations and made their way farther up the placer. However, when they encroached on Erel's licensed territory, many were evicted by armed security guards and local police.[7] Throughout my fieldwork in Uyanga, ninjas continued to mine for gold on Ölt, which was desired for its relatively high concentration of gold and at its peak hosted approximately seven thousand ninjas. Since the deposits in some areas ran at a depth of sixteen to eighteen meters below the surface, it was grueling work to dig the deep mining shafts using only handmade tools such as small metal picks (*skov*) and hammers (*zeetüü*). As Odgerel, a forty-year-old herder, who had arrived in the mines in 2001, commented despairingly, "Before you dig the hole, you don't know whether or not there will be *any* gold at all. There might be nothing."

Once the mining shaft (*nüh*, lit. hole or opening) revealed growing concentrations of gold, ninjas branched out and dug horizontal tunnels (*tünel*) following the deposit. They removed the mined deposit and all the unwanted dirt (known as "overburden," *nabor*) from inside the tunnels, often in buckets and bags that were hauled to the surface by means of a simple hand pulley (*libotok*) (see figure 7). The pulley was operated

Figure 7. Ninjas relaxing by a mining hole and waiting to haul more bags to the surface. Credit: Jos Sances, Alliance Graphics.

by two people working on the ground above. Since mining shafts and tunnels were not reinforced, there was a high risk of collapsing mines and landslides. Many ninjas therefore moved into the neighboring valley called Shar Suvag (lit. Yellow Vein), hosting approximately one thousand ninjas in the summer. This placer had a lower concentration of gold, but the buried deposits were much shallower, running at a depth of only about six meters. During the long winter, when temperatures dropped below minus forty degrees Celsius and mining became nearly impossible, the remaining ninjas gathered in Shar Suvag. In order to soften the frozen ground, they set fire to dried yak dung and rubber tires, which burned slowly inside mining holes. They used blowtorches to melt the ice and to keep the water from freezing, techniques that were relatively costly, time-consuming, and detrimental to both the environment and the ninjas' health. Although most people joined the gold rush in the warmer summer months, mining thus continued incessantly throughout the year.

The Long *Zud*

In order to understand why so many local herders got involved in this physically demanding and highly dangerous work, it is important to emphasize from the outset that zud decimated millions of animals during three consecutive winters from 1999 to 2002—that is, the very same period that the gold rush gathered momentum in Uyanga. This period was locally referred to as *udaan zud*, meaning a long, slow, and lingering zud. There are many forms of zud, such as *tsagaan* (white), *har* (black), *tömör* (iron), and *hüiten* (cold). Some result from high snowfall, which prevents animals from reaching the grass below. Others occur when there is a lack of snow, thus leaving animals with very limited water supplies. Still others happen when the snow cover melts and refreezes, thereby creating an impenetrable iron-like ice cover that encases the grass so the animals cannot graze. All these various zuds can cause high animal mortality, and the zuds of 1999–2002 were particularly severe. In fact, they were the worst that Mongolian herders had experienced in sixty years (Batima, Natsagdorj, and Batnasan 2008:77). During the three years, more than twelve million livestock died, representing about 25 percent of the country's total livestock population (National Statistical Office of Mongolia 2003). Fortunately, the winter of 2003 proved less harsh, and the fertile summer of 2004 marked a definite end to the long zud.

During the 1999–2002 cycle of disaster, thousands of families, especially in the Gobi desert region, lost their entire herds, and many others were suddenly pushed into poverty. This mass death of livestock affected many herders' ability to sustain a life on the steppe, and its effect could be felt in Uyanga's mines. When I initially tried to gather basic census data, I came upon a large number of former herders from the Gobi region. Many of them told me about their livestock losses and desperate search for a new subsistence base. Although they found the work conditions in Ölt and Shar Suvag difficult, they felt that there was no longer a possibility of making a living in their own *nutag* (homeland). Echoing the views of many, a thirty-three-year-old mother of two said, "Our nutag has nothing left for us anymore. Life is now very difficult." Ninjas from other parts of Mongolia, including Uyanga, told me about similar experiences of sudden and abject poverty. They told me about their frantic attempts to buy fodder and improve animal shelters. They described how they fought for their animals' lives by taking many of the weakest yearlings into their gers to keep them warm. Yet many still saw their livestock starve to death, and they recounted detailed memories of how carcasses were strewn across the steppe. The disastrous zud left a trail of misery, which contributed to many people's move to Ölt and Shar Suvag.

However, my fieldwork revealed that the gold rush also attracted people from wealthy households. Interestingly, of the ninjas who came from Uyanga and its neighboring districts (*sum*), it seemed that most were from herding households. Moreover, they were often from households that had *not* suffered major livestock losses during the zud of 1999–2002. Like Ahaa, these ninjas included some of the wealthiest herding households, the so-called number one households (*negdügeer ail*) that earned their informal titles from the enormous size of their herds. It is thus clear that macroeconomic indicators of poverty and deprivation do not alone explain the wider attraction of mining. The loss of herds was certainly central to many people's motivation for joining the gold rush, but it does not account for the involvement of the many wealthy local herders. Taking cues from people's own concerns, in the rest of this chapter I will consider the emergence of a mining economy from a perspective that includes attention to not only the dynamics of wealth creation but also the organization of family life on the steppe—a way of life that gave rise to much frustration and discontent.

Partings and Returns

As one arrives in the district of Uyanga, the low-lying steppe land starts to rise tall, forests appear, and distances between ails increase. Generally, in September the winter sets in and lasts through April. At this time of year, ails usually have the fewest household members. Some children will be in the village, where they stay with relatives and attend school. Others might have gone to the capital to go to university or look for a job. And again others might spend time in the mines, especially when Tsagaan Sar (the expensive lunar new year celebrations, lit. White Moon) approaches. Since herders' animals are hardly lactating, there is not much work to do on the steppe in the wintertime, and many hours are instead spent sitting around the stove, drinking tea and chatting. The winter diet is usually meager, consisting of meat, animal fat, and homemade noodles. Dairy products from the previous summer are kept cold or frozen and carefully rationed through the barren season. Around the month of April, the quiet and strenuous monotony of winter comes to an abrupt end as the lambing season sets in, requiring as many helping hands as possible. Ninja miners, schoolgoers, and elderly relatives might return to their families to help out. As the summer approaches, ails relocate to new pasture in order to provide the best conditions for the herd in terms of grass, water, and shelter. If ails have too many animals for the immediate pasture, some of the family members may take part of the herd with them and join a smaller ail for a couple of months. During the lush summer months, the herd rapidly fattens up and yields great amounts of milk. The hitherto meager diet is now supplemented by a variety of fresh dairy products, such as milk, fermented mare's milk (*airag*), cheese, yogurt, milk curd (*aruul*), butter, and clotted cream (*öröm*) made from yak, mare, sheep, and goat's milk. The village shops have a burgeoning stock of imported vegetables, such as potatoes, onions, carrots, cabbages, and turnips, which can be bought at relatively low prices. However, like many others in the region, my hosts preferred meals without vegetables and included them only for very special, celebratory occasions. During the summer, many adults and children return from afar and spend a relaxing summer holiday with herding relatives. Benefiting from the warm summer days, many also go to the mines between June and September before the frost again sets in and renders panning for gold extremely demanding. When the first snow falls, usually in mid-September, the weakest animals are taken out on an extended migration (*otor*) to fertile pasture in order to better prepare them for the coming winter. Some of the nicely fattened animals are slaughtered and then sold, perhaps in the village or in the mines. The autumn slaughter helps cover the costs of winter fodder (*hiveg*, lit. bran), bought at the bustling market in the provincial capital of Arvaiheer several hours' drive away. As the winter preparations begin, ails have again sent off relatives and incorporated those who wish to spend the next lonesome winter together.

These changes in the size and composition of ails depend not only on the season but also on major family events like marriage (see also Cooper 1995). For in Uyanga marriage is predominantly virilocal, which means that the wife leaves her natal home and moves in with her husband's ail. When the wedding of one of Yagaanövgön's daughters was approaching, she was both elated and sad. Joining her fiancé's family, who lived in the neighboring district, she told me her worries about not being able to see her family again for a long time. She knew that it would be entirely at the discretion of her future husband and his father to let her visit her nutag, and, in her view, they were unlikely to support such visits. Indeed, her own mother had never been back to her nutag since getting married more than forty years earlier. Her nutag was only sixty kilometers away, but Yagaanövgön, and his father when he was alive, had not let her visit. Not even once. Heads of households told me that they thought it was best if affinal households resided relatively far from each other. According to Yagaanövgön, it was "much easier" (*mash ih amarhan*) that way because otherwise "daughters-in-law would want to visit their families all the time." This preference is encapsulated in the following proverb, which I often heard reiterated by household heads:

> If water and snow are close together, it is good.
> If relatives-by-marriage are far apart, it is good.
> (Vreeland [1954] 1962, 64)[8]

When Yagaanövgön's daughter married, she moved with her husband into their new ger in his father's ail. They tended to her dowry herd (*injiin mal*), as well as his inherited herd (*huviin mal*).[9] Some household heads choose to divide their herds into equal shares (*huv'*) and pass on the animals to their sons and daughters before their death. Others prefer for the inheritance to wait until after they have passed away. The youngest son usually stays with his parents and inherits both his share of animals and all of the parents' material property—gers, cars, and motorbikes as well as smaller items such as stoves, saddles, buckets, and the like. The other sons might remain in their parents' ail for a time or join ails of other agnatic kin, a decision that often hinges on issues such as the size of their herd, the quality of the pastureland, and the social relationships that they have forged (see also Vreeland [1954] 1962, 79–88).

There is a noticeable emphasis on agnatic kin ties within and beyond ails in Uyanga. This residential pattern might be a long-standing regional feature. In a study of herders' residential groups in the 1930s in the neighboring province of Arhangai, Simukov (1933, 24) notes that "there is a clear tendency to join up in *hotons* [ails] according to kinship lines" where "agnatic kinship between household heads was the most important" (quoted in Sneath 2000, 214). While this stress on agnatic ties might be characteristic of the local area, the elevated position of male household heads has been described as part of a so-called Mongolian episteme (Pedersen 2001, 419)—a notion that is intended to convey the specific historical conditions of possibility (Descola 1996; see also Foucault 1970). In his bold comparative study of North Asian indigenous societies and ritual practices, Morten Pedersen (2001, 420) argues that in contrast to northern Siberia, Mongolia and bordering regions are dominated by "vertically" organized social formations that are produced and reproduced "through notions of inherited leadership, a hierarchical ethos, patrilineal descent . . . , etc." For Pedersen, this general vertical organization of social life means that a given man "will be bound up in a hierarchy of patrilineal clans delineating for him a clearcut territory to act within, in spatial, social, and existential terms" (ibid., 420). This patriarchal structure with its clear delineations appears "as if someone laid a grid over the world" (ibid., 418; see also Buyandelger 2013, 34, and Humphrey 1995). Verticality, in the sense of hierarchical difference, is thus a fundamental characteristic that informs how life is lived in this region.

However, within these vertically organized social formations, it seems that the position of Uyanga's male household heads is significantly more authoritarian than that described for other parts of Mongolia. Writing about Halh Mongols in western Mongolia, Vreeland ([1954] 1962, 54) for example notes,

> By 1920 the average senior woman of the family appears to have been almost on an equal footing with her husband. She had something to say in almost any issue. . . . She could take money from the family funds, on her own initiative, to pay for a temple service for some special occasion, and she could veto, or at least protest, expenditures by her husband or any man who was the nominal trustee for the family property. Her agreement had to be secured in any divisions or allotments of family animals.

In Uyanga, male household heads hold autonomous decision-making power and consult other household members before making a decision only if they choose to do so. These decisions concern all aspects of daily life, ranging from quotidian herding matters to extraordinary ritual events. A case in point is Yagaanövgön's denying his wife participation in the commemorative ritual that was held one year after the death of her mother. Despite her tearful anger and firm insistence, he would not allow her to make the journey to the Buddhist monastery in the village. Their oldest daughter was also facing the decision-making power of her husband. Following a clash between him and his father, they had moved to the village, where they lived on their own but struggled to get by. By September 2005 she had fallen seriously ill and had to go to hospital. Relatives stopped by and gave them money to enable her to travel to the provincial capital and receive medical treatment. However, her husband decided to spend the money himself, and her medical state deteriorated steadily throughout my fieldwork.

Given the extent to which household heads in Uyanga often single-handedly decide matters that may at times severely affect other members, it is perhaps not surprising that people occasionally hold strong emotions of resentment and anger. Although such disagreement is often initially suppressed, as described in the introductory vignette to this chapter, there were times when overt violence erupted (see also Delaplace 2009). The following excerpt from my field notes describes a dispute that took place shortly after my arrival in Uyanga, involving Nyambuu, the father of my first host family, and his oldest son, Bayasgalan.

> Earlier in the day, Bayasgalan had arrived from the mines with his wife and six-year-old son. Food was prepared and people were chatting, laughing and having a good time together. In the evening Nyambuu came back from a day of herding and drinking. He took a seat in the northern part of the ger, received a large serving of food, and hungrily started eating. Bayasgalan then received his portion and soon everybody turned their attention to the food. But halfway through the meal, Bayasgalan asked his dad why the timber he had paid him to buy hadn't arrived yet. His dad gave a long, windy explanation and then continued eating. Bayasgalan stopped eating and looked angrily at his dad. "You liar! You haven't bought the timber, have you?!" *Ber* [her common name, lit. daughter-in-law] immediately interjected: "Don't say that!" But Bayasgalan continued: "I should go myself and get it!" His wife begged him to calm down, but Bayasgalan jumped up and threw his porcelain bowl with full force at his wife and mother. They ducked just in time for the bowl to crash to pieces against the kitchen shelves behind them. Bayasgalan tried to catch hold of his wife and hit her, but Nyambuu ran across and intervened. Bayasgalan tried to hit her again and then stormed out of the ger and drove off.

In this situation Ber became the target in a dispute for which Bayasgalan and Nyambuu were responsible. As a loyal in-marrying wife, she defended her father-in-law. But her own submissive position within the household also invited Bayasgalan to take his anger out on her rather than on his father. Given the position of Nyambuu within the ail, frustrations with his actions thus exposed the hierarchical relations among the other family members.

Writing about contemporary Inner Mongolia, David Sneath (2000) describes much more balanced gender relations than I observed in Uyanga. He notes, for example, how daily decision making is not carried out autonomously by the male household head but is rather a process of consultation between husband, wife, and children. Moreover, compared with that in Uyanga, the gendered division of labor is noticeably less strict. Sneath (180) writes,

> In practice there is a large amount of overlap between male and female work. If they are without the company of those of the opposite gender, men and women have to cope with both

> types of work. Men cook when there are no women around (such as when they are working far from their ail) or if their wives are ill. If they are in the household men will often provide some level of childcare, and sometimes help prepare food—particularly when the family treats itself to *bansh* [small meat dumplings] which the whole family usually makes together.

Across the regional literature, there are numerous examples of a much more careful coexistence of quasi-egalitarian and hierarchical relations within pastoral domestic economies (see Empson 2011; Humphrey 1998, 287–99; Sneath 1993; see also Benwell 2006 and Dalaibuyan 2012, 48–50 for a discussion of gender relations in urban areas). This coexistence might correlate with greater variation in postmarital residence, greater proximity between affinal households, more balanced interaction with both matrilateral and patrilateral kin, and/or earlier division of inheritance.[10] I am not postulating a simple causal relationship between local kinship practices and their hierarchical organization in daily life. Instead I am suggesting that gendered hierarchies have a particularly strong foundation in Uyanga. Local practices, such as virilocal marriage and distant matrilateral kin, certainly offer little challenge to the singular position of household heads like Yagaanövgön and Nyambuu.

The Predicament of the *Ber*

According to Caroline Humphrey (1993b), it was a long-standing tradition in the Mongolian cultural region that men married women of higher status, a practice that is also known as "hypogamy" and is a structurally highly unstable arrangement (see also Lévi-Strauss 1969, 241). The instability arose because of potential tensions between a man's ties to his wife's kin versus his ties to his own kin—ties that could pull in conflicting directions. If a man instead married a woman of lower status, her kin could, in structural terms, be either ignored or used as clients. In this way, ties to her kin would not pose a similar threat to the structural continuity of the wife-taking group. In Mongolia, marriage with higher-status women was part of a wider constellation of cultural values, integrating women into both material and religious aspects of the pastoral domestic economy (Atwood 2004, 314).[11] As is reflected in myths and epics, wealthy fathers were not easily lured by young men's bride-wealth propositions for their daughters, and as a result intense and fearful bridal trials were a necessary recourse for marriageable men (see, for example, Hamayon 1987). The men had to subordinate themselves through the trials of acquiring a wife, but the wife was also accompanied by a dowry (*inj*), which remained her personal property. According to Humphrey (1983b, 184), "the basic idea which is conveyed by the Ch'ing period law codes is that an adult woman should have rights to enough productive property to be able to live autonomously if widowed or divorced."

Given the structural instability of hypogamy and the practical tribulations through which Mongolian men acquired their wives, Humphrey concludes that, once married, the higher-status wife necessarily became subject to intense regimes of subordination. It was only through her subordination that her husband's kin group could reproduce itself without her posing constant threats to its perpetuity. The subordination involved an unparalleled long and hard working day, carrying out the same tasks over and over again, leaving her little freedom to do anything else (Sokolewicz 1977). Also, as depicted in a still-cherished Mongol epic, the father-in-law would present her with riddles to test her wit and cunningness, thereby making her act according to *his* terms of discursive interaction (Hamayon and Bassanoff 1973). Yet in her everyday life, what required constant adaptation and loyal compliance was the necessity for her to conform to the extensive taboo (*tseer*) on her language, which transformed her speech markedly from everyone else's. The daughter-in-law was "absolutely forbidden to use the names, either in address or reference, of her *xadamud* (her husband's older brothers, his father, his father's brothers, grandfather, etc.). The taboo includes the names of the wives of close *xadamud*. The *ber* is also prohibited from mentioning the name of her husband's patriclan. Furthermore, she is strictly enjoined not to use any word in ordinary language which enters any of the forbidden names or sounds like them" (Humphrey 1978, 91–92).

As the ber replaced the words and sounds of the patriclan members, her personalized speech reflected her wit but also ensured that she did not attract the attention of the person named, especially her father-in-law. Such attention was undesired within the kin group. Bringing with her a sizable dowry, the ber could potentially return to her natal family and even bring along her husband and children. Her threat to the father-in-law's household was thus not so much grounded in her own ambiguous feelings as in her capacity to tempt her husband and patriclan descendants away. Since the inheritance of property went from father to son, a son was likely to grow impatient waiting for the transfer of leadership from brother to brother within the agnatic group after receiving his inheritance share (*huv'*). If he invested in his nuclear family, conjugal sentiments were a direct challenge to the continuing authority of his kin group. The seed of the patriclan's destruction was therefore contained within its own structure, in which the transfer of leaders was at odds with the transfer of property. Conflicts between brothers fill Mongolian epics and accounted for even the infamous dissolution of the Mongol Empire. In order to avoid such a fate, the in-marrying ber with her tempting autonomy was marginalized verbally and ideologically through practices such as the extensive language taboo.

On the basis of her research among present-day Buryats of northern Mongolia, Rebecca Empson (2011) shows that daughters-in-law can still today be seen to be in a "state of liminality" defined by their recurring possibilities of separation and transformation. Whereas public rites integrate and transform women into loyal daughters-in-law, tensions occasionally mount and achieve full expression through the actions of the ber. In her analysis of a rebellious daughter-in-law who decides to instigate temporary separation from her husband's kin group, Empson (2003, 77) shows that a ber can "subtly *subvert* certain rules and maintain her position as not fully part of her husband's family while outwardly seeming to comply with her expected behaviour as a daughter-in-law" (emphasis in original). This subversive potential among daughters-in-law was also evident in Uyanga, perhaps most explicitly in Yagaanövgön's household.

Ahaa's wife was not subject to constant riddles or extensive language taboos. Yet she was still considered an ambiguous and indeed potentially disloyal member of the ail. Her marriage to Ahaa had been arranged eight years earlier by Yagaanövgön, who had been close friends with her father. The two men used to drink heavily together, and Yagaanövgön had much respect for her father. This respect (*hündetgel*), he explained, was sometimes bordering fear (*emeeh*) since her father was a renowned black magic specialist (*har zügiin lam*, lit. lama of black direction). People requested his help in casting curses (*har haraal hiih*), and it was important for Yagaanövgön to stay on good terms with him (*oir dotno hariltsaatai*). To arrange a marriage between their children seemed to him to make good sense. One summer my host family needed help with their large herd, and Yagaanövgön requested the black magic specialist's daughter to come and help with the milking duties. Ber and Ahaa soon fell in love and received their fathers' approval quickly.[12] Her family had been a relatively well-off herding family, but shortly before she got married, her parents started to lose their animals to the wolves and harsh winters. A few years after the wedding, their herd was decimated, and they were forced to relocate to the mines in search of a new livelihood. Her father then died, and Ber lost two children prematurely. It was a period of constant grief. Shortly after giving birth to her first surviving child, she was so frustrated with her subordinate position within Yagaanövgön's ail that she left for the mines with her son. Her mother and oldest brother were in Ölt, and she could stay with them. A long process of negotiation started, whereby Ahaa joined her in the mines and his father paid visits, begging them to return. In private conversations, members of my host family told me that they believed Yagaanövgön wanted Ber to return partly because of her presumed insights into malevolent ritual practices. It was crucial to convince her to return and continue peaceful conviviality. She did eventually return to the ail and is still living there, despite frequent threats of leaving for the mines.

Although rows between Ber and Ahaa have become part of everyday life, the following excerpt from my field notes describes her own views of the situation. A row erupted after Ahaa had bought a small storage ger from an acquaintance of his father and had paid much more than he expected. When he unpacked the ger, everyone was shocked to see its poor condition. He had been cheated, and his wife reacted by beating him hysterically and shouting furiously in front of the entire ail. The situation was awkward for all parties. I went by their ger later in the day to see if she was all right.

> Ber was alone in the ger when I entered. She passed me a cup of tea and said, "I can't believe he bought such garbage [*hog*]. It's useless! And he paid 70,000 tögrög and two sheep! I told him to wait with any payment until I could see the ger too, but he didn't listen to me. Why? Because he's just like his dad, always deciding everything on his own, thinking that he knows better [*ööröö ilüü sain medne gej bodoj baina*]. He always thinks he knows better! He won't listen to me. You know, here with my parents-in law, my father-in law decides everything himself and my mother-in law is so complacent. She accepts whatever he decides, she puts up with anything [*tevchij baina*]. She is so modest [*daruuhan*] and humble [*nomhon*]. If they expect me to be like that, I just can't [*chadahgüi*]!

As Ber struggled to take up the expected subordinate position within Ahaa's ail, she often threated to return to the mines. In that case, her husband (the first in line to inherit the leadership after his aging father) *either* had to dissolve his nuclear family and let go of his wife and son *or* separate from his father's ail and let down the patriclan that he had always been expected to carry on. Despite the daily hardship within the ail, Ber had a greater degree of autonomy and leverage than did other members of the household group. Since she had Ahaa's son, her dowry herd, and a continuing emotional tie with her natal family, her in-laws were forced to subtly negotiate her subordination lest she leave and go to Ölt.

Tensions within the *Ail*

It was not only daughters-in law who felt that the way life was organized in Uyanga's herding households was a burden. Although Yagaanövgön often declared that he did not want any of his children to become ninjas and instead wanted them all to remain herders like himself, I suggest that the daily experience of patriarchy can ultimately push some children into mining. The hierarchical relations within herding ails are not separate from or irrelevant to the emerging mining economy but instead are central to the local appeal of the gold rush.

Baajiimaa, Yagaanövgön's oldest unmarried daughter, was twenty-six years old and anxious to move away from the ail. She had once started seeing a local young herder in secret, and although she told me about his bad temper and taste for alcohol, she had entertained the idea of marrying him and moving away. However, one day Yagaanövgön discovered her with her boyfriend (*naiz zaluu*) and he became furious. In front of others, he made the young man promise never to see his daughter again. Baajiimaa was sad and angry, frustrated by her father's constant interfering in what she perceived to be her own matters (*ööriin hereg*). Disagreements and suppressed arguments seemed to constantly accompany interactions between father and daughter, especially when he made suggestions for possible husbands or when she expressed interest in men other than those her father preferred. The following excerpt from my field notes concerns one such situation:

> Baajiimaa and her dad had argued earlier in the day. She got upset and stormed out of the ger. The rest of the day she kept to herself. I later walked down to the river, when Baajiimaa appeared. She walked toward me. When she reached me, she closed her eyes and brought her hands to her face, covering her eyes. She said, "I sometimes feel really depressed [*zarimdaa bi setgeleer undag*]." I asked what troubled her. She looked at me. "I don't know . . . all kinds of things. I just feel sad [*uitgartai*] and worried [*sanaa zovoj baina*]. I think a lot about my life, what I want, what I could have done differently, what if I had gone to school, how would my life then look? Dad always decides [*shiideh*] everything and doesn't listen to anyone. It doesn't matter what I say, he never listens to me. He doesn't care about me [*toohgüi*]." I interjected, "But he doesn't listen to *anyone*, not even Mum or Ahaa!" Baajiimaa replied, "That's true, true. But it's still annoying that he decides everything and it's always all on his own. If I want to do something, he tells me that I can't [*chadahgüi*], that I mustn't [*bolohgüi*]. How will he ever start listening to me, respecting me, include me in decisions? It just makes me so upset that Dad always decides everything himself."

Yagaanövgön had made one decision in particular that frequently upset the equanimity of his children: they had not been allowed to go to school. Out of the ten children, only the oldest son, Ahaa, had attended the local school for a couple of years before Yagaanövgön decided to pull him out. Although frustrated about not having been offered more book-based learning, all of my host siblings seemed more preoccupied by their isolation from meeting potential spouses. Because they lived in a remote mountainous region, the village school was the nexus for meeting nonrelatives. Yet in the turbulent time of 1990s, when the seventy-year-old socialist regime collapsed, Yagaanövgön decided to sacrifice his children's enrollment in school so that they could continue their life as herders. Ever since, he had not allowed any of his children to go back to school. Although the many children were obvious assets to the running of a herding household, there was not always much work for everybody to do. As the younger children begged their father to be allowed to attend school, their requests were met with a wall of silence and disregard. As the ultimate authority on the allowed acquisition of skills and social relations, Yagaanövgön did not engage with their pleadings.

Many other herders living in the region took a similar stance, and as a result the village school had reputedly one of the highest rates of left-out pupils in the country. It was very difficult to obtain reliable, if any, statistical data on this issue since many residents in the district of Uyanga were not registered with the local administration. Moreover, the local schoolteachers were reluctant to discuss this matter as they were supposed to provide long-distance teaching to the left-out pupils through home visits. However, these visits seemed to be more official rhetoric than actual practice. In Mongolia, primary and secondary school education is officially free, and dormitories are meant to be available at a low cost for children from herding households. But in practice Uyanga's dormitories were so ill maintained that children could not stay there during the winter when the teaching took place. Furthermore, with high levels of corruption, education was an expensive investment for parents.[13] Yet another reason for the numerous left-out children was the existence of a close link between the local school and ninja mining.

In Mongolia, all public-sector jobs—whether for teachers, doctors, dentists, or policemen—offer low official salaries. At the time of my fieldwork, Uyanga's secondary school teachers earned approximately 80,000 MNT (Mongolian national currency) per month (67 USD). During the school year, many teachers took their pupils on daily excursions to Ölt and Shar Suvag. They arrived in small minibuses, and the teachers had often planned in advance to have some ready-to-mine dirt available. Depending on their personal networks, this was sometimes arranged with an Erel employee or a resident ninja. The children then squatted by the panning pools and washed gold alongside their teachers. After a few hours, they all returned to the village school. When I met them by chance in the mines, both pupils and teachers described what they considered a mutually advantageous arrangement: part of the pupils' earnings was given to the teacher, thus augmenting his or her salary. The other part they were allowed to keep, thus earning good pocket money for themselves.

When Yagaanövgön and other household heads decided not to enroll their children in the village school, it was not necessarily the distanced, cold calculation it often seemed to those affected. Given the school's proximity to and connection with Ölt and Shar Suvag, many worried that if their children attended school, they would perhaps move to the mines and become ninjas. They would perhaps begin to follow gold rather than the herds. The ability of household heads to prevent their children receiving state education is not only an indication of the striking power they hold over other household members. It is also an extension of their common, and possibly growing, role in arranging marriages. This is because the village school is associated with book-based learning as well as meeting potential marriage partners—it is seen as a place where love relationships can happen. In a region where many herding children are denied school education and expected to marry the spouse of their father's choice, the school has become an evocative symbol of individual freedom, the freedom to find your own spouse and decide on your own livelihood. However, for Yagaanövgön and other household heads it is associated with the breakdown of herding society, its patriclans, and loyal youths. It is a symbol of how a way of life is coming to an end.

A classical theme in the regional literature has been how relatively balanced and cooperative relations within nomadic pastoralist households in Mongolia enable people to respond to the constantly changing needs of their large herds. As this chapter has demonstrated, however, in Uyanga the dynamics within herding households differ markedly. I have shown how, in line with deeper historical trends, pronounced patriarchal relations predominate in this particular part of the country. And I suggest that the efforts to sustain household hierarchies, more than being merely a remnant of the past, are intensifying with the advent of the gold rush. If we abandon the notion that nomadic pastoralists in Uyanga or elsewhere constitute a naturally bounded social whole that is tied to its ecology and economy, the proximity to and presence of the gold rush can be seen as part of the ongoing dynamics of herding life. Rather than constituting the "acculturation" of a previously isolated or pristine group, the attraction of the gold rush is part of today's nomadic pastoralism in Uyanga and its long-standing social hierarchies. In this way, the tensions and despair that I have described in this chapter reveal the patriarchal foundations and their transformations in the context of contemporary translocal processes.

Recognizing the importance of the gold rush for pastoralist practices in Uyanga, however, does not mean that all ninjas are driven by a temptation or desire to part from their herding households. This is only part of the story. In the following chapter I consider how a nonhuman substance known as the power of gold (*altny chadvar*) is now also deemed central to the perpetuation of the mining boom. With a capacity for social intervention, the substance can affect people's ability to mobilize action, and it invites us to begin apprehending how the gold rush is far from an exclusively human affair.

2 The Power of Gold

WHEN I FIRST met Ganbat, he was in his midthirties and lived on the steppe with his elderly widowed mother, Enhjargal, and their four yaks—a tiny herd that did not produce enough dairy and meat for their daily sustenance. He had seven siblings, so his mother should have enjoyed a comfortable old age in which her sons carried out the hard herding tasks and their wives took care of the domestic chores, as described in the previous chapter. But all her sons remained unmarried and were still a burden on her household. Ganbat himself had developed a strong liking for hard alcohol and put Enhjargal under major economic and social strain. To help her out, Nyambuu, who was her patrilateral cousin, had invited them to stay with his *ail* during the summer. He then herded her animals with his and generously shared his flour, rice, and tea, while in return she and Ganbat helped with chores such as shearing goats and sheep, mucking out animal shelters, and drawing water from the nearby river. Whereas Nyambuu's *ger* was beautifully decorated with commissioned furniture and luxury items, Enhjargal's small ger had no electricity, table, or stools. In Enhjargal's presence, Nyambuu's wife often reminded me unnecessarily that "Enhjargal is poor" (*Enhjargal yaduu baina*).

Over the course of the summer, Ganbat mentioned that he wanted to work at the bottom of a mining hole. His younger sister and some of his brothers already worked in the mines, and he had visited them on a few occasions. "I want to direct the tube!" he said enthusiastically. The tube (*hooloi*) was a flexible pipe attached to a pump above ground that sucked up gravel from deep inside the dug tunnels. "Directing the tube" was known to be one of the most dangerous jobs in the mines with the highest risk of accidents. One morning Ganbat was nowhere to be seen and did not return all day. As the day drew to a close, Enhjargal concluded that her son had probably left for the mines, and she asked people to look out for him. After Ganbat had been away for more than a month without sending any message home, Enhjargal started to grow increasingly concerned. I went with her to collect firewood one day and as we caught our breaths after the steep walk with the heavy baskets on our backs, Enhjargal shared her thoughts: "My son is a good person, but sometimes people treat him badly. They hit him and tease him. I just don't know how he is doing. I'm worried about him [*tüünd sanaa zovoj baina*]." With no money, Enhjargal could not afford to travel the twenty

Figure 8. Offering a piece of *aruul*

kilometers to the mines, where most of her children were now making a living. "He is vulnerable [*emzeg*]," she continued. "He can so easily be pulled [*tatagdah*]."[1] "By the vodka?" I asked. "No, by *altny chadvar*." She handed me a piece of *aruul* (dried milk curd, see figure 8) and motioned for us to resume our walk back to the ail, which could be seen in the far distance.

When I bumped into Ganbat in the mines about a year later, his bruised face lit up in a smile, revealing several gaps from recently lost teeth. "I still work in the holes [*nüh*]," he said proudly. "And I'm still fighting!" He laughed while punching and apparently knocking out an invisible foe. He emotionally asked about his mother, uncle, and other relatives. Brushing some caked mud off his dirty T-shirt, he told me how much he wanted to return to the steppe. "When will you go?" I asked, thinking about the many evenings when his mother had been distraught by her concern for his well-being. "I don't know . . ."

The transformation that has happened in Uyanga is not simply willed or wanted—even by those who have actively decided to join the gold rush. In the previous chapter, I illustrated the desire to follow gold (*alt dagah*) rather than animals in contexts that involved discontent with the dominant power of household heads or the sudden reality of abject poverty following the long *zud*. These, however, are not the only possibilities. Whereas some scholars have carefully traced how people seek, and to a large extent appear to manage, to influence the course of their own futures (e.g., Sahlins 1985, 1992; Robbins 2004), I discuss here a situation in which radical transformation is not accompanied by similar manifestations of human sovereignty and self-determination. Indeed, alongside human agents, a nonhuman substance is regarded as central to the advent and perpetuation of the mining boom. With scores of ninjas feeling unable to fully determine their own actions, I approach the gold rush as a cosmoeconomy within which humans and nonhumans occupy the same landscape and interact with the same precious metal. These continuities demand an analytical approach that does not reduce capacities such as agency, intention, and representation to necessarily those of humans only (Kohn 2013, 42). How other kinds of beings or things act in the world matter. But rather than getting carried away by the excitement of the theoretical, analytical, if not also ethical, potential in this affordance of unfamiliarity, I draw on and limit myself to the cosmological evocations advanced by my interlocutors in Uyanga. Attending to this coexistence of diverse beings, in this chapter I show how altny chadvar has become part of a transformation that is at once desired and disliked, embraced and rejected by its human participants.

Washing Dirt

In Uyanga's mines, thousands of ninjas live in tattered gers and camping tents pitched closely up against each other (figure 9). Keeping pace with the overturning of the land and the creation of new mining holes, ninjas are constantly searching for suitable living areas. With mining continuing day and night, the appearance of an area changes radically within weeks. Ninjas divert rivers, move excavation areas, and turn hillsides into living quarters. Nothing remains permanent, least of all the landscape that is excavated at great speed. In this transient environment, ninjas often reorganize themselves in relatively stable residential groups, comprising three to six gers or tents, each housing about five people. Within these clusters that are referred to as ails, just as on the steppe, people assist each other in daily household tasks and work-related activities. From early morning, kitchen utensils, cooking ingredients, and mining equipment are passed between gers. Outside men are often hunched on their heels in a circle around somebody repairing his motorbike, sluice box, or diesel generator. A quick sneak onto company grounds may have revealed a new mining technique that could be tried out. Or a visiting miner from the north may be entertaining the crowd with fascinating stories from the distant mines. Packets of cigarettes and bottles of vodka circulate, and banter, if not fighting, fills the air.

Figure 9. Residential clusters in the mines

These ninja ails provide the basis for mining teams in which men and women, young and old work alongside each other. In order to work on a mining team, people do not have to enter into binding agreements, pay monetary deposits, or agree to any other kind of formal arrangement. Newcomers, including the visiting anthropologist, are easily accepted onto the teams and can quickly learn how to use the mining equipment. No machine is regarded as particularly female-operated or male-operated, or as particularly appropriate for younger or elder workers. This allows mining teams to be highly flexible and accommodating. As one of my ninja hosts said, "We are good at what we do. All we need in order to do our work is our hands." This bodily emphasis is also reflected in the official Mongolian term for artisanal mining, which is *ashigt maltmal gar üildverleliin argaar olborloh*, literally meaning "mining for minerals by use of hands." In colloquial language, people in Uyanga usually refer to artisanal mining as *shoroo ugaah*, which translates as "to wash dirt." In line with the experiences of many other ninjas, a friend detailed how his mining team was organized. He explained,

> On my team we are thirty workers, divided over three shifts [*eelj*] of twelve hours. So the work is like this: Morning to evening shift [*ödört*] with ten of us working for twelve hours, then the next day I have the evening to morning shift [*shönöd*] and then I get the third day off. On the team we are a total of fifteen men and fifteen women. It just happens to be so that we have the same number of men and women. It's not always like that. It just depends . . . whoever is a good worker, man or woman, whoever can do the job well can work on the team. It doesn't matter where you are from and whether you are a relative of someone on the team. All that matters is that you are hard-working and reliable. If you don't turn up for one of your shifts, the others on the team will get really upset. . . . They will probably beat you up, severely [laughter].

When people arrive in the mines in search of work, their ninja relatives usually welcome them by showing lavish hospitality. A quick run to the ger shop (*geriin delgüür*) ensures that bottles of Korean beer and Mongolian vodka are bought. A visit to the butcher ger ensures that fresh mutton can make the prepared meal special. And someone is quickly sent to the mining hole to fetch the ninja relative. Apart from often leading to a drinking binge, these reunions also commence the process of finding a suitable mining team for the visitor. Drawing on their networks, they often try to get their relatives accepted onto someone else's mining team rather than their own. I asked them why there was this reluctance to work together with their own relatives, but I rarely received answers other than *medehgüi*, "I don't know." Once newcomers have found a mining team, they are quickly incorporated into their own residential clusters and end up socializing more with their own friends than with their relatives. Beyond the occasional visits and drinking binges, the presence of relatives in the mines is therefore barely noticeable in everyday life.

Depending on the location and size of the mining hole, mining teams number four to thirty people with some working above ground and others inside the mining holes. If the mining shaft is far from the panning pools, so-called dirt taxis (*shoroony taksi*) are paid to take the bags of dirt from the mine to the water. My neighbor, Byamsüren, was one such dirt taxi driver. He had worked for Erel as an *uurhaich* (miner) for more than ten years and had saved up some money. When the gold rush started, he eyed a lucrative opportunity. He bought himself a Russian jeep and launched his new dirt taxi business. In his view, this was much better in terms of money and safety. Yet his wife worried about him. "He works such long hours. And with the rains and hails, it gets dangerous. He is just driving on dirt with deep holes underneath. . . . It's dangerous. I don't like it."

For every load he transports, Byamsüren receives 5,000 MNT (4.15 USD), and he can do many runs over the course of his workday, which stretches from dawn until dusk. If people want to save the cost of the dirt taxi, they sometimes simply carry the bags to the panning pools. As described in the previous chapter, there are various techniques ranging from using simple panning pans and rubber mats to diesel-powered generators and large sluice boxes (see figure 10). Once they have retrieved the tiny specks of gold, ninjas carefully wrap up their findings and take it to the locally

Figure 10. A mining team using a sluice box to separate and retrieve the gold. Credit: Jos Sances, Alliance Graphics.

based *altny chanj* (unregistered petty gold traders) to sell it for cash at the end of each workday (see also chapters 4 and 6). From my fieldwork among both ninjas and gold traders, I estimate that most ninjas earn a minimum of 5,000 MNT (4.15 USD) per day. If they use more elaborate mining techniques, their earnings easily grow manyfold. In a country where the national average income at the time was 41,000 MNT per month (29 USD), ninja mining thus offers a very attractive income comparable to that of white-collar workers such as legislators, senior officials, managers, and other professionals (National Statistical Office of Mongolia 2006, 108).

To my surprise, all members of a mining team receive an equal share of the day's earnings regardless of their age, gender, or experience. Even people working inside mining holes and directing the tube receive the same remuneration as those working above ground. This is despite the fact that they are facing much greater dangers such as landslides, cave-ins, and poisonous gases. Bilgee from my mining team explained that it would be impossible not to divide the money equally in an area that is so extreme (*ontsgoi*). Given the daily fear and reality of violence, unequal remuneration would likely give rise to physical retribution and immediate exclusion from the area. As she put it, "It is easier this way. Otherwise people will just get angry." For those who live in an area as densely populated as the mines, where labor regimes require people to work so closely together to carry out often dangerous tasks, unspoken rules underlie social interactions and are strongly enforced when violated.

Considering the predominance of hierarchical relations in Uyanga as described in the previous chapter, the organization of mining teams and the remuneration of labor place an extraordinary emphasis on the ninja as an equal individual, independently of personal connections, skills, or contributions (see High 2012). This emphasis on the equal value of labor is striking and, I suspect, unprecedented (cf. Vreeland [1954] 1962). As Caroline Humphrey has shown in her earlier work (1979, 1983), even during the socialist period, when ideologies of solidarity and policies on equality were espoused in the country, the actual organization of labor remained largely patriarchal and gerontocratic. Gender differences provided the basis for many regimes of hierarchy, which were often reinforced through seniority. In Uyanga's mines today, such hierarchies are downplayed and made irrelevant to how labor is organized, and they are further downplayed in ninjas' domestic life. It was striking for me to experience how the conventional ordering of space does not apply in the mines. Rather than regarding certain spaces of a ger as honorable, hosts and guests, young and old, male and female intermingle freely. Also, rather than beginning with an extensive formal prologue, conversations just flow at once. And, most strikingly, the common verb forms *yostoi* (have to) and *heregtei* (need to), which any visitor to Mongolia will hear repeatedly, are not used in the mines. These verbs presuppose the existence of a hierarchical relationship between speaker and listener, implying that the speaker is in a position to command the respect necessary for imposing obligations on others. These expressions are common among herders when directing household members who are in a lower social position than the speaker. But it is only within such social hierarchies that speakers can expect listeners to accept their demands. In the mines, such expressions are largely limited to situations of emergency, and requests for assistance are instead expressed in collective voluntary terms, conveying the speaker's wish rather than demand (*-ya, -ye, -yo,* let's . . .). The collective verb form presents the suggestion as markedly nonobligatory, expressing and producing through discourse the kind of relationship that they are taking part in.

One would perhaps think that in a place like the mines, hierarchies would inevitably develop when a ninja made a major gold discovery. Nugget finds, as opposed to small particles of gold, do indeed occur in Uyanga, albeit very rarely. When a ninja discovers a gold nugget, it does not become a collective find of the mining team. It becomes his own and acquires his name when remembered and talked about years later. When someone finds a nugget, an elaborate and overindulgent celebration known as *bühiin bayar* (celebration of the nugget) follows. These celebrations are famous for the excessive hospitality that is being shown by the lucky ninja. Everybody is invited to join the celebrations. There is no sense of demarcation or boundary making with regard to who can take part. As one of my ninja hosts said, "If someone finds a nugget, everybody gets lucky!" Just before one of my stays in the mines, a nugget worth apparently more than 34 million MNT (28,333 USD) had been found. Even with a population of approximately seven thousand people, most mining teams stopped working, and people enjoyed the supply of great food and unlimited alcohol. It was only when all the funds were spent that the celebration stopped. Apart from buying himself a Chinese motorbike, the ninja who had discovered the nugget had spent all his money. Given these celebrations, nugget discoveries do not really lead to noticeable wealth accumulation. As a result, unlike in the herding households on the steppe, in mining communities there are no "number one" ninjas who have accumulated large amounts of wealth, whether be it in mining machinery, cash savings, or indebted laborers. The processes of hierarchization and formalization, so central to many gold rushes elsewhere in the world, have at least not yet emerged in Uyanga. Instead it is a place where the value of individual autonomy influences and dominates the organization of daily life (see also High 2016a).

Mongolia's New Nomads

In economic history, gold rushes are often singled out for their particular macroeconomic dynamics of expanded money supply (Frost 2010; Rawls 1999). As miners discover gold, they rapidly bring more money into circulation. For Milton Friedman and Anna Jacobson Schwartz (1963), gold rushes were fascinating economic phenomena because they illustrated what happened when the supply of money was not kept steady. Across the world, the sudden and definite peaks brought about by major gold discoveries presented the fastest velocity of money in times of peace. Examining the consequences of such irregular money supply, they saw that, on the one hand, gold rushes were a warning against central banks' tinkering with money supply while, on the other, they affirmed the financial potential in large-scale entrepreneurship. Gold rushes became "synonymous with opportunity and the potential to achieve anything"

(Walsh 2005, 6) as miners were seen to conquer new and risky spaces for wealth creation (see also Jønsson and Bryceson 2009). Considering the nonhierarchical organization of labor and remuneration in Uyanga, this potent and persuasive frontier imagery of the "optimizing individual" might seem to also capture what is going on in Mongolia's gold mines. However, many ninjas feel that their actions are not quite so controlled and self-authored. Introducing some of Uyanga's ninjas and their personal journeys to the mines, I will show how the gold rush offers a new way of life that is both enticing and troubling, even for those who have decided to become ninjas.

Dalai and Tsegii are a married couple in their early thirties. They are from Darhan-Uul, an industrialized region in northern Mongolia where many mining companies operate. They first heard of Uyanga's gold from some of their friends in Zaamar, which is the other major ninja mining hub in the country, located a four-hour drive from Darhan. Their friends had visited Uyanga every summer over the previous few years and always returned with mesmerizing stories about the huge deposits of high-quality gold. "Our friends made it clear that this was where to come for 'yellow stuff' [*shar yum*].[2] And they were right!" Dalai recounted. "My wife and I have an apartment in Darhan city, but we had no work. So we thought, why not come here?" They had never tried their hand at mining before and were a little apprehensive at first. How were they going to mine? What did they need to bring? What was it going to be like? They had heard many stories about life in the mines, few of them positive.

Following the advice of their friends, they started panning farther upriver in Ölt, where there was more gold to be found. But the police, who officially claim to confiscate illegal gold and evict ninjas from company grounds, caught them by surprise several times. One day, when they were washing gold in the river, they suddenly heard the sound of an approaching Russian minivan. Dalai and Tsegii quickly packed away their panning bowls (*tümpen*) and rubber mats (*erzeen*) and began running in the direction of their ger. They ran as fast as they could, but the minivan hurled past them and applied the brakes, and a swarm of policemen jumped out. Dalai and Tsegii threw their things on the ground and tried to escape but in vain. The police surrounded Dalai and took turns beating him up while Tsegii shouted for them to stop. The next day Dalai and Tsegii decided to pack up their belongings and leave. Their novice experiences of ninja mining had been upsetting and they wanted to go back to Darhan. However, when they tried to arrange a car, someone persuaded them to try the quieter area farther downstream, where I met them in 2006.

Having found a new place for their ger, Dalai and Tsegii visited the nearby *mergen hün* (fortune-teller). Dorsüren was a short, somewhat corpulent woman in her early sixties. On my first visit to her ger, she told me that her father and father-in law had both been lamas prior to the religious purges in the 1930s. They had introduced her to Buddhist texts, the divination calendar, and nine-coin divination (Swancutt 2012, 79). "I'm not someone who knows a lot and can do much for other people. I just sit here, on my bed, in this shop." She pointed around her at the boxes full of dried pasta, rice, and noodles, then piles of panning bowls and rubber mats, and again more boxes containing cheap vodka and cartons of cigarettes. Running her business out of the corner of a ger shop, she continued, "If someone has a problem, if someone wants me to help them, I will try. But I'm not like the accomplished [*dadlagatai*] lamas of Uyanga." Dorsüren was from Arvaiheer, a three-hour drive away, where she had worked in a state-owned store. As soon as she and her husband heard the rumors about Uyanga's gold, they packed their belongings and moved. "That was in 2001! And we have been here ever since. Work is going very well," she said, laughing heartily. Approaching her as someone who knew about "invisible things" (*haragdahgüi yum*), Dalai and Tsegii wanted her to shed light on two things: first, whether it was a good idea to stay in the mines and second, whether she could see anything that was pulling at them. From her coin divination, Dorsüren concluded that they should stay and that there was probably no altny chadvar affecting them. "But be careful," Dorsüren warned them. "Up there you were just washing dirt. Down here there are deep holes. You were already worried about altny chadvar? You should be more worried here!"

Dalai and Tsegii were quite pleased with the amount of gold they had found so far. With their earnings they bought a large television, a DVD player, and stacks of DVDs. They put a handwritten sign outside their ger, announcing that it was now a "movie ger" (*kino*), where for a small fee, people could enter and watch films. It was a good business, and Tsegii was spending more time in the ger tending to customers than doing what she considered dirty (*hirtei*) and heavy (*hünd*) mining work. Most of her customers were children, and they reminded her of how much she wanted to one day start a family of her own. Since she did not like the idea of bringing up a child in the mines, family life for her ideally involved a move back to their apartment in Darhan. But Dalai was not so keen on giving up mining. He had just bought a dry sluice box from another ninja. No longer dependent on the use of water to separate gold from the gravel, Dalai joked that they had become Mongolia's new nomads (*shine nüüdelchid*), capable of taking on even the Gobi Desert! As autumn approached, rumors of great gold discoveries upstream began to circulate. Someone had found a nugget, an event that was becoming increasingly rare in Uyanga. Dalai was enthralled by the prospect of striking it rich, but Tsegii was a little concerned. Were they ever going to leave the mines? Was altny chadvar at work? We soon bid our farewells, and they moved back up to the area where they had lived before.

Dalai and Tsegii were repeatedly drawn to the possibilities they saw in Uyanga's mining camps, and a return to their former way of life in Darhan seemed less and less likely. Like many others in the mines, they envisioned futures that increasingly centered on mining opportunities. They had learned how to mine and how to do it well. They had also had much success with their movie ger and had dreams for potential future businesses. After only a few years in the mines, they had come to accept the dangers and embrace the excitement of "following gold." As a ninja told me, when you wake up in the morning, you don't know how the day will end. Each day is distinct and unpredictable, sometimes for the better and sometimes for the worse. Guiding them through the unpredictability and showing them a brave way of living, Dalai and Tsegii venerated the fearless tunnel diggers who risked their lives for the prosperity of the mining team, and they took their cues from the few prominent ninjas whose extraordinary luck (*az*) had led them to make singular discoveries. In referring to themselves as ninjas, Dalai and Tsegii did not merely identify with the social role of ninjas in the sense of having acquired the necessary technical skills and having become good at gold mining. They also organized and oriented their lives in ways that drew centrally on these new and emerging ideals for how to live well.

Altny Chadvar

Although people actively join the gold rush, many ninjas do not feel similarly capable of leaving the mines when they want to. Many find that their ability to leave is then suddenly restricted. When casual conversation touched upon this, many ninjas began to describe the gold rush as a kind of centripetal force that pulled (*tatah*) them in. "We can't move away from the mines," a ninja on my mining team commented when she and her husband failed to move back to the steppe. "Every time we try, we are pulled back [*tatagdah*]." She made a fist and pulled it strongly in a direct line toward her body. Another ninja, who had previously mined for gold in the neighboring province of Bayanhongor, noted that "there is some kind of strong thing [*hüchtei yum*] here in Ölt. It wasn't in Bayanhongor. It was very different there." Some relatives in the village and on the steppe described ninjas as *chadal bagatai*, meaning "having little strength." Degidsüren, the wife of Nyambuu, even described them as "powerless" (*hüchgüi*), incapable of doing what they intended to do. This severance of action from intention is expressed as a manifestation of so-called altny chadvar: the power of gold. As previously mentioned, this is an invisible and potent substance that emerges when people extract the heavy and dangerous (*ayuultai*) gold from the landscape. Given the invisibility of altny chadvar, its materiality is little known, but some people described it as a sort of invisible dust (*haragdahgüi shoroo*) that rubs off gold during the process of extraction. It can be transferred onto others through physical contact or reputedly manipulated through cursing rituals. In the next chapter I will consider in more detail the cosmological understandings that underpin the emergence of such forces. For now, it suffices to note that the released altny chadvar is said to cause a particular illness (*övchin*) among its victims: an obsessive longing (*hüceh*) or greed (*shunahai*) for the precious metal.

People do not know exactly if and when they have become afflicted by altny chadvar. There is rarely a clear and distinct moment that can be identified as the so-called moment of affliction (Eliade [1951] 2004, 29–31; Lewis 2003, 54). There is nothing akin to the "inner metamorphosis" that neophyte shamans are said to experience. In her work among the Daur Mongols of Inner Mongolia, Caroline Humphrey (1996, 31) describes how a young man or woman would fall ill, experience disorientation, and suffer both mentally and physically. This incurable state culminates in a terrible vision in which "one might see oneself as dying, as being eaten by spirits, or taken apart limb by limb, and being forced by the spirits to accept a shaman ancestor spirit (*onggor*). . . . Reconstituted physically by the spirits, or at least acknowledging that the suffering was spirit-caused, and consciously accepting the *onggor*, the initiate recovered and took on a new existence as a shaman" (see also Pedersen 2011, 109).

In Uyanga, there is no such dramatic or recognizable moment that reveals the effect of altny chadvar on the person. There is also no way of measuring the presence or absence of the invisible force. In a fascinating account, Katherine Swancutt (2012, 116–23) details how the Buryats of far eastern Mongolia can gauge whether a person's soul is present. According to them, people with a lost soul will find that one ring finger has become slightly longer than the other. There was no single explanation for this, but some people reasoned that it was because the soul left the body through the ring finger (cf. Humphrey 1996, 255n56). Using a spoon handle, a pencil, or any other straight-edged item, the measurer simply compares the lengths of the person's ring fingers. Without these overt indications of metamorphosis, people in Uyanga are much less certain about whether altny chadvar really has afflicted them. It is a constant possibility that threatens people's affective state. Consequently, the search for gold is not presumed to be necessarily willed by a sovereign human subject. Instead, it potentially evidences the power and presence of an invisible, nonhuman substance.

As Marilyn Strathern (1997, 142) observes, actions that are carried out unconsciously—for example, under the influence of magic and sorcery—are often not subject to the same moral evaluation as conscious actions. In this sense, unconscious actions may be regarded as amoral rather than immoral. As ninjas join the very industry that is making local life difficult, the potential presence of altny chadvar unsettles any quick and easy denunciation of their actions. Blurring distinctions between human and nonhuman agency, we can see how people's actions are the manifestation not only of their human contexts but indeed also of that which lies beyond. Rather than regarding mining as a clear act of human volition, we see that altny chadvar is partly responsible for what has happened in the area in the past and what can happen in the future.

As an agent of transformation, altny chadvar is an important temporal marker. People in Uyanga generally agree that there was no altny chadvar prior to Erel's operations, or indeed before the Soviet geologists' visits. When I suggested that maybe a valley was so rich in gold that it somehow naturally released such forces, my proposition was met with outright disagreement. "Absolutely not! [*yestoi ügüi*]," a ninja exclaimed. "It isn't like that." Another maintained, "It has to be released [*gargagdah*]. It's only when people do something that it emerges." Positioning human agents behind the release of altny chadvar, people insist that the current circumstances have a knowable and identifiable culprit, namely, the owners of Erel. Although there is little agreement as to the exact time of its first emergence, they maintain that there was no altny chadvar that could possibly have compelled the company owners to have carried out their gold-centered work when they arrived in the early 1990s. The exploratory work that had been carried out by the Soviet geologists since 1940 had definitely not contributed to the release of altny chadvar because, people reasoned, the geologists had only taken small samples, not initiated large-scale mining. I carefully suggested that maybe the company owners had not known about altny chadvar, and if they had not, maybe they could not be blamed for the later emergence of the gold rush. But people insisted that it made no difference to the effects of their actions. Since the owners of Erel alone had set in motion the forces that eventually led to the radical transformations in Uyanga, the moral blame was placed directly on them.

The Battle for Water

The conjunction between gold's extraction and an invisible substance evokes a cosmos that is centrally implicated in Mongolia's mining boom (High 2013). However, according to some ninjas, altny chadvar is not present in all of Mongolia's gold mines. Indeed, it seems that it is a rather rare substance that is potentially unique to Uyanga. There are other parts of Mongolia where there are also large gold mines but apparently no altny chadvar. For example, the "biggest mining exploration project in the world" (Grainger 2003), the Oyu Tolgoi copper and gold mine in the South Gobi province, is described as safe (*ayuulgüi*) from altny chadvar. Similarly, Zaamar appears largely free of the substance. There was little certainty as to why altny chadvar had not emerged in these other mines, many of which the ninjas had personally visited and worked in. Acknowledging that the invisible force is not simply logically related to scale or technology and is often unfamiliar to people from other parts of the country, I will here examine how the local concern with altny chadvar and the condemnation of Erel has been negotiated in national politics and popular media, far beyond the immediate context of Uyanga.

The staunchest opponent to Erel is without doubt Tsetsgeegiin Münhbayar, the founder of the Ongi River Movement (Ongi Golynhon Hödölgöön) (Byambajav 2015; Upton 2012). I first met Münhbayar in his office in a dilapidated building block in Ulaanbaatar in January 2005. He told me that he was from a family of camel herders in the South Gobi province and that one day he had begun to notice that the Ongi River was drying up. As the then chair of the Local Citizens' Council (Irgediin Tölöölögchdiin Hural), he organized several meetings in the nearest village and confronted provincial government officials with people's concerns over water scarcity. He tried to obtain environmental impact assessment reports for the mining companies operating along the river. But although these documents are public records according to Mongolian law, the Ministry of Nature and Environment refused to release them. Surprised by the apparent reluctance among senior politicians to heed the concerns of local people, Münhbayar began to galvanize support for his environmental movement. Together with dozens of new Ongi River Movement members, he staged a protest in 2002 involving hundreds of camels at Sühbaatar Square, right in front of the parliament building in Ulaanbaatar. Attracting much public attention and the support of international development organizations, the Ongi River Movement brought petitions signed by thousands of people to the prime minister, who eventually ordered the Ministry of Nature and Environment to conduct research into the causes of the river's depletion. In a typically passionate speech during a protest march along the river in 2004, Münhbayar was recorded saying, "We are just a small group of simple herders fighting powerful people. I know this is not an easy fight but we can't stand by idly and watch our land and way of life come to an end."

In our many private conversations, Münhbayar lamented the continuing release of altny chadvar in the region and insisted that Erel and other mining companies immediately cease their destructive activities and begin rehabilitating the land. In his view, that was the only way altny chadvar could ever be hindered or delayed (*saatuulah*). The holes had to be filled with soil, the dams had to be opened, and the tailings had to be leveled. *Baigal'* (the landscape) had to be cleaned from the presence of mining (*tseverleh*) and made pure once again. Only then would the moral madness of the *altny hiirhel* (gold rush) be tamed. Because of altny chadvar, people were now out of control and wanted gold so badly that they did not care about anything or anyone else. The hiirhel had to be stopped, and, in Münhbayar's view, only someone from outside Uyanga, someone who was positioned far away from the release of altny chadvar, could do this. Someone like members of parliament.

Münhbayar was relentless in his efforts, and eventually a parliamentary commission conducted the promised research in 2007. Followed closely by the public, the inquiry investigated several mining sites, including Erel's operations in Ölt. In an interview published in a widely circulated Mongolian newspaper, a member of the commission laid bare the conclusion that Erel had been purposely forgoing the environmental rehabilitation procedures mandated by Mongolian law. He stated that "the responsibility for all the mess created in Uyanga sum should rest with Erel Company" (*Ödriin Sonin* 2007). Several members of parliament began to openly attack Erel for the situation in Uyanga. Then a debate followed, broadcast on the state-funded TV network MNB (Mongolian National Broadcasting). The owners of Erel declined to participate, and in their absence, the debate placed the responsibility for the "ecological catastrophe," as the State Property Committee chairman, Zandaahüügiin Enhbold, described it, squarely on Erel's shoulders. The national media focused exclusively on the mining company, rarely mentioning the thousands of ninjas who were also mining for gold along the Ongi River.[3] Declaring Erel the sole perpetrator was not only a potentially fruitful legal strategy but also an attempt among politicians to wage their own political battles. The owner of Erel, Badarchiin Erdenebat, in 1998 had founded the Motherland Party (Eh Oron Nam), popularly known as the Erel Party.[4] In consecutive elections the party won seats in parliament, and eventually the owner of Erel became the minister for fuel and resources, and the CEO of Erel became the minister of nature and environment.[5] The Erel Party thus came to occupy some of the most influential ministries for policies and investigations directly related to mining. But these ministerial positions lasted only two years. The exposé revealed not only the company's bad practice but also its obvious conflicting interests. Consequently, Erel officially ceased its operations in Uyanga in 2007. Since the politicians had fought against Erel for their own political reasons, it is perhaps not surprising that the environmental concerns raised by Münhbayar and the Ongi River Movement retreated into the background. To date, Erel has neither been prosecuted for its environmental misconduct nor been forced to rehabilitate the land and the river.

Although Erel was rumored to still be profiting from the mines in Uyanga by having subcontracted its operations to other companies, Münhbayar was applauded internationally for his fight against Erel and thirty-five other mining companies operating in the Ongi River Basin. He was celebrated for having "convinced the government to increase and enforce mining regulations in the region and to stop damaging mining activities and begin environmental restoration." Such were the words when he was awarded the National Geographic Emerging Explorer title in 2008, and when Münhbayar received the Goldman Environmental Prize of 2007. Layton Croft, vice president of Ivanhoe Mines—which operates the enormous Oyu Tolgoi mine in Münhbayar's own province—commented that "the key to Münhbayar's success as a leader for responsible mining in Mongolia is that he has had the courage to acknowledge that mining could be good for Mongolia, as long as it is done in a very open and participatory way."[6] As a figure of international renown, Münhbayar found himself becoming a valuable commodity for not only politicians but also the mining companies that he was fighting against. The very people he opposed co-opted his agenda, and he became their key icon of Mongolia's grassroots consent to mining.

In an increasingly desperate battle, Münhbayar was joined by members of other environmental organizations in a desperate hunger strike on Sühbaatar Square in 2009. At the time it was still legal to mine near rivers, and they wanted the parliament to ratify a new law to protect rivers from the effects of mining. However, the pro-mining government did not change its stance on environmental regulation, and Münhbayar, together with the United River and Lake Movement, attempted to sue it for the "environmental harm done to eight river basins as a result of the government's inactivity and failure to uphold the norms set forth in the Constitution" (*Novaya Gazeta* 2011). His fight was no longer just about the Ongi River but about all of Mongolia. When the court dismissed all charges, Münhbayar declared his readiness to use violence against environmentally destructive mining companies. After his alleged participation in a shooting that targeted mining equipment in Uyanga, he told the press, "We will give the mining companies fair warning—either they must cease their activities or incur our wrath. If they do not comply with our demands, then we will use our guns. We are not violent people but we will do what we need to do to stop these environmental polluters" (Kohn 2011).

Since receiving his international accolades for convincing the government to increase environmental regulation, he has seen seventy-eight new mining licenses issued in Övörhangai province alone (MRAM 2012, 42). "People have criticized our choice [to take up arms] but, tell me, what could we hope to achieve through peaceful means?" one of his coactivists said (http://www.news.mn, September 16, 2010). In the words of another coactivist, "The most salient question for Mongolians today is not whether mining should occur here. There's no point [to that question], because it is happening anyway. . . . We fought for eight to ten years to stop mining companies, and it doesn't happen. Why? Because it happens with or without you" (Awehali 2011).

As Münhbayar and fellow activitists become increasingly marginalized from political debates, they exemplify both the scale of the mining boom and the difficulty of opposing it. Processes of radical transformation are sweeping across Mongolia, but even as Münhbayar can see that "our way of life is threatened" (Lovgren 2008), his fight for the Ongi River and against Erel has been largely unsuccessful. Although the battle for water has coalesced around a single perpetrator against whom ninjas, herders, and others could unite, Münhbayar appears just as powerless as anyone else in Uyanga. Maybe, if I asked him today, he would say that altny chadvar has now spread to even Ulaanbaatar.

In his work on the Azande, Edward Evans-Pritchard (1937, 513) famously noted, "New situations demand new magic." This insight has survived the classic structural-functionalist approaches and has been reaffirmed repeatedly within later theoretical frameworks in anthropology, be it "spirits of resistance" (Ong 1987; Boddy 1989), "the modernity of witchcraft" (Geschiere 1997); "occult economies" (Comaroff and Comaroff 1999), or latest "spiritual economies" (Rudnyckyj 2011). These various bodies of work are concerned with the junctures between radical economic change and seemingly esoteric phenomena like altny chadvar. Although we could easily fit altny chadvar into the same analytical straitjacket and claim that the invisible dust is an a posteriori attempt at sense making, if not an evocative symbol of affliction, this chapter has shown us that the people in Uyanga see it differently. For they approach the agentive dust as a directly relevant, indeed fundamental, participant in Uyanga's gold mines. Rather than explaining altny chadvar as simply a response to or an effect of hegemonic structures, I have focused on its existence as an integral part of people's lives (see also Pedersen 2011, 34). Affecting people's ability to mobilize action, it emerges as a constitutive component of the gold rush, central to its emergence and proliferation.

But altny chadvar is not the only nonhuman agent that takes part in human life. As the following chapter shows, multiple spirit beings also inhabit the same landscape, especially the mines, where new and even more powerful beings are said to have emerged. Through this exploration of spirit worlds, mining will emerge as a cosmologically distinctive and dangerous act.

3 Angered Spirits

A HEAVY HAILSTORM hit Uyanga at the height of summer in 2006. Hailstones the size of coins pounded on the ground, and everybody ran to seek cover. The noise escalated and within minutes the area was covered in a thick layer of ice. Hunched by the slightly open door, Yaanjilham, Yagaanövgön's wife, whispered to herself, "What is this? How long will this last? What about our poor lambs? And Tömörchödör, he is still out herding the sheep. . . . Why hasn't he come back yet? Maybe something has happened . . ."

She got up, filled a ladle with fresh milk, and hurried outside. She threw the milk toward the sky, mumbled some words, and hurried back in. She repeated the action twice before the thundering hail finally began to subside a little. The sky slowly cleared, and her son eventually appeared. He was bruised badly by the hail and had fallen off his horse. A few minutes later one of his sisters arrived, carrying a shivering and bleeding lamb well-protected inside her *deel* (robe). I was handed the lamb and asked if I could keep her warm. After an hour, the lamb was getting warmer, but it was still not certain that she would survive. Yagaanövgön lit a butter candle on the household altar, and I asked him what had caused the bad weather. He replied, "It was *tenger* [the Sky] who wanted to wash the ground, the animals, everything. By having thirty minutes of hail, all the filth [*buzar*] and garbage [*hog*] has been pushed away, leaving the ground clean [*tsever*] and peaceful [*taivan*]. Basically tenger was cleaning up everything. It's very important once or twice a year."

A few kilometers away, in the heart of the mining areas, the hailstorm had hit even harder. When I went a few days later, I could see how the outer fabric of most *gers* (*geriin bürees*) was torn to pieces, and the usual few makeshift plastic tents were completely destroyed. "Hail the size of eggs came down here!" Bayarmaa, an elderly female ninja from my mining team, recounted. "It even shattered the windshields of cars! Can you imagine?! It was awful!" Several others agreed: "Tenger sure was angry! [*Tenger uurlasan shüü*]."[1] In the weeks following the hailstorm, both herders and ninjas referred to the hailstorm as an important warning (*sanuulga*) that tenger did not approve of the mining activities in Ölt and Shar Suvag. As Yagaanövgön reasoned, "That is why the hail was so much bigger in the mines! Tenger took its anger out on ninjas because they refuse to show respect [*hündetgehgüi*]."

In this chapter I examine how the changing interactions with the landscape entailed in herding and mining are conceived within local cosmology. My primary concern is with the multiple, and sometimes mutually incongruent, understandings that people have of the relations between humans and spirit beings. Thus my aim is not to attempt to distill a singular, internally coherent worldview that is predicated on assumptions of statis and holism (Humphrey and Onon 1996, 76; Bubandt and Otto 2011, 11). Nor am I interested in establishing an ahistorical cosmology that is held by an exotic Other (Abramson and Holbraad 2012, 36). Instead, by attending to how herders and miners in Uyanga regard the various spirit beings with whom they interact, I am interested in cosmologies as active and dynamic constituents of economic life. As spirits roam across and inhabit the landscape, they are centrally involved in both the pastoral economy and the extractive industry of mining. Many scholars have noted that in mining areas around the world spirit beings are often described as epiphenomena of human attention to fundamental taboos (see, for example, Clark 1993; Nash 1979; Taussig 1980; Walsh 2006). This is also the case in Mongolia. However, spirits are also central to how interhuman relationships are constituted and enacted. They matter for the position of household heads, described in chapter 1. And they matter for the social distance desired by ninjas described in chapter 2. When we recognize the expansive and inclusive world within which people act, economic life emerges as a potent manifestation of human and nonhuman interaction.

Living with Spirit Beings

The regional literature offers many examples of the various kinds of air-like (*hii yum*), invisible (*haragdaggüi*) beings that humans interact with (see Buyandelger 2013, 16; Humphrey and Onon 1996; Pedersen, Empson, and Humphrey 2007; Swancutt 2012, 66–71). Some of these beings are considered particularly, if not exclusively, malevolent for humans. In the mountains of Uyanga, there are the ever-hungry beings known as *berd* whose throats are said to be as long and thin as a needle, thus unable to swallow any food. In a state of desperate but insatiable hunger, berd violently take their frustrations out on those people who happen to cross their paths. And there are the wandering ghosts called *shulmus* who take great pleasure in scaring, or attacking, travelers at night. Hiding behind rocks or riding on untamed horses, shulmus are tricksters of the dark who excel at making sudden and unexpected appearances. Very rarely, people also encounter the much-feared *almas*, who frighten them to such an extent that they subsequently resign themselves to withering or even dying. All these beings are said to have a troubled human origin and are directed by human-like vices and desires. Yet despite their close familiarity with the conditions of human life, their interference in human life is not predictable or even knowable. There are no particular actions that are known to upset these beings, nor are there particular rituals that are known to appease them. Their appearance in and influence on human life are menacing and beyond human manipulation.

People in Uyanga distinguished these unpredictable invisible beings from another category, which has become emblematic of north Asian religion. These are *ongon* (pl. *ongod*) or "shamanic spirits." However, the region of central Mongolia is not known for its shamanic practice. As a villager remarked, "The people of Uyanga have no *udha*," referring to shamanic abilities. This apparent absence of shamanic specialists means not only that there are no recognized shamans but also that there is no evident "shamanic landscape" (Humphrey 1995). Writing about the Darhad of northern Mongolia, Morten Pedersen (2011, 166) outlines how souls of deceased shamans over time turn into ongod, which are absorbed into particular features of the landscape like trees, rivers, and mountains. Ongod are in this sense part of the history of past shamans (see also Buyandelger 2013). However, the current absence of shamanic specialists in Uyanga need not preempt the future emergence of udha in the area. Among the Darhad, udha skips generations before manifesting itself again (Pedersen 2011, 99). So while it is possible that ongod might one day acquire a presence in Uyanga, at the moment they are not considered significant in daily human life.

A third category, which is recognized to have a fundamental presence and role in everyday life in Uyanga, consists of spirit beings known as *gazryn ezed* (masters of the land), *usan khan* (water lords), *savdag* (local nature spirits), and *lus* (head deities of the landscape). All these beings occupy specific domains of the landscape, such as forests, rivers, and mountaintops (Humphrey and Onon 1996, 76). They usually have no body, gender, or color, and lack souls (*süns*). Whereas humans have individual souls that require attention and care during lifetime and are passed on to others at death through rebirth, these spirits do not have such temporalizing and affective characteristics (cf. Kristensen 2007, 277). They are omniscient and immortal, capable of seeing and hearing everything that humans do. Talking about gazryn ezed, Ahaa put it this way: "They can see everything you do and they know everything you do. They are like politicians where the governors of *bags*, *sums* and *aimags* all talk with the politicians in Ulaanbaatar.[2] In this way politicians know everything. That's also how it is with gazryn ezed."

In everyday practice and conversation, these spirit beings are circumscribed by elaborate prescriptions and encompassing ideas. It is with regard to them that practices such as herding and mining are usually discussed locally. Since these spirit beings are seen to reside in and even "own" particular parts of the landscape, it is hardly surprising that activities such as penetrating the surface of the land, unearthing the subsoil, and panning in streams are seen to upset them more than they would the roaming ghosts or faraway shamanic ongod.

There are numerous taboos (*tseer*) that prohibit specific forms of human interaction with the landscape. During my fieldwork I was informed about these taboos whenever I unknowingly or forgetfully failed to observe them. At such times people were quick to point out my mistakes and the impending danger I had caused. On the steppe (figure 11), herders reminded me most often of the following taboos:

- You must not dig into the ground (*gazart uhaj bolohgüi*).
- You must not break off fresh branches from trees (*noiton mod avch bolohgüi*).
- You must not put anything dirty in rivers, streams, or lakes (*usand bohir yum orch bolohgüi*).
- You must not break stones (*urgaa chuluug hovhloj bolohgüi*).

- You must not break off the small twigs of berry bushes (*jimsnii ish avch bolohgüi*).
- You must not collect wild garlic (*zerleg songino avch bolohgüi*).
- Women must not climb up in pine trees (*emegtei hün hushand avirch bolohgüi*).

At times I felt frustrated by the seemingly endless list of prohibitions on my interaction with the landscape. At such moments I asked people for explanations of why certain behavior was seen as transgressive. Most often they just looked at me and sighed, providing no further explanation. But occasionally I received short statements such as this: "Trees are alive [*mod am'tai*].[3] You should therefore always collect only the dead wood on the ground. If you break off a fresh branch, even by accident, you hurt the tree. It doesn't like that, so it will get upset at you. Maybe it is not that particular tree that will get upset at you but *hangain lus* [the mountain *lus*] protecting the forest. If lus gets upset [*uurlaval*], it is very bad for you and your *ail*."

Recognizing the landscape as having life, feelings, and agency, these taboos instruct people to avoid activities that deny the presence of other beings in their own right.

Figure 11. The landscape of the steppe

Nature spirits become upset, and masters of the land cause illnesses and other adversities if taboos are not adhered to. Agency and intention are thus not considered a human monopoly, but rather a shared feature of the interrelated forces that collectively constitute *baigal'*. The term *baigal'* refers to the existence of nature, commonly translated as "environment," and derives from the verb *baih* (to be). This verb form is used to describe characteristics of humans, animals, and material objects, and it conveys the multiplicity and diversity of beings that make up the all-encompassing baigal'. As long as each being suppresses its disturbing autonomy in the system of interrelated forces, a balance between powers is achieved, and, for people, living becomes wonderful (*saihan*) and peaceful (*taivan*). Statements such as "we and our pastures are one body" (Yenhu 1996, 20) emphasize this practical and cosmological interconnection, if not dependency, between humans and others. Indeed, "if one part of nature denies the existence of another then eventually it will be denying its own" (Tseren 1996, 147). The conditions for existence are relational and premised on affording others respect.

When people fail to show such respect—for example, by transgressing taboos—they often conceive of later unfortunate events as manifestations of spirits' anger. For example, a friend told me of a woman who once climbed up in a pine tree to pick nuts and shortly afterward the branch broke off. The woman hit the ground hard and broke her hand. The branch broke and she got injured because she ignored savdag, my friend reasoned. Also, one of my host sisters once went to collect firewood, but instead of taking the dry branches from the ground, she broke off fresh branches. As she mounted her usually calm horse, it suddenly jumped high up and threw her off. She injured herself badly and today ascribes this to her mistake of breaking off fresh branches. Numerous stories such as these circulate. On the steppe as well as in the mines, people narrate them among themselves when sitting around the stove or sharing a bottle of vodka. These narratives about broken taboos provide compelling entertainment but also serve to remind people of the importance, and indeed difficulty, of interacting morally with the landscape (see also Basso 1996; Walsh 2006)—that is, of outlining how one ought to act.

In Uyanga the spirits are considered moody and temperamental (*aashtai*). As a result, it is impossible to predict the form their anger will take or its precise timing. Since the original transgression and the subsequent repercussions may be separated by weeks, if not months, an extended period may pass during which people contemplate whether a particular unfortunate occurrence really was due to angered spirits or whether further misfortune is still to come. Spirits are thus often represented in daily discourse as anticipated agents that will eventually manifest themselves in human misfortune. But spirits are referred to not only as a way to rationalize misfortune but also as beings that are present in the landscape independently of human action. Transgressions of taboos are not necessary for people to start talking about spirits, as the following example illustrates. It was narrated by Yaanjilham and concerns the common genre of spirit warnings (*sanuulga*). We had just taken our evening meal and were waiting for the herd to return from pasture when she told me this:

> One evening we had the door of the ger open and suddenly three completely white things [*tsav tsagaan yum*] appeared. They never entered the ger but instead remained outside. We didn't know what they were. They weren't stones or birds or anything we had ever seen before. We all got really scared [*aisan*] and just waited for them to go away. As soon as they had left, we did readings [*unshlaga*]. I think it was definitely lus warning us.

Such spirit warnings are not rare, and it seems that many people in the area have experienced them at least once. It is not a particularly new phenomenon, people told me. Spirits have warned humans in this area for a long time, also before the gold rush. Rather than marking a fundamental change in human-spirit relationships, this narrative genre testifies to the local presence and importance of spirits. They also highlight the extensive and creative potential of spirits by which they can momentarily take on bodies, genders, and colors. They can also rely on other beings to pass on their warnings. This connection between beings was brought to the fore in an account narrated by one of my host sisters:

> I was riding my horse near the cliffs along the dry riverbed. Suddenly, out of nowhere, two huge birds appeared with their eyes glowing red, staring right at me. They were big. And I think they were black. They were definitely dark. They were huge. The birds came closer and closer and began picking at me. I made my horse gallop as fast as possible away from the birds. I just wanted to get away! I have never before or since seen such birds but I'm sure it was lus warning me.

At first when people narrated such spirit warnings, I reacted by asking if someone had done something wrong, such as failing to observe particular taboos. What exactly was lus warning people against? In these situations I received puzzled looks and the simple reply "Don't know." I soon realized that these narratives were centrally concerned with spirits rather than humans. Whereas I had initially focused on human actions as necessitating these warnings, other listeners requested more details about the spirits. With regard to the completely white things, for example, the others present asked, "Were they big?" "How white were they?" "Did they stay for long?" And when my host sister talked about the birds with glowing red eyes, she did not need my prompting in order to want to describe them in more detail. In talking about spirits, people seemed to attempt to turn the invisible into concrete and tangible beings.[4] The warnings were in this sense not about particular human acts but rather concerned with the incontestable presence of spirits.

Frequently reminded of such forces, people generally try to respect spirit beings. By consulting the lunar calendar, divinatory coins, and Buddhist lamas, they sought advice regarding which days would prove benign to carry out what they considered disruptive actions. If permission had not been granted (*ödör garig tseerleh,* lit. to make a day taboo), people were likely to either try another divinatory technique or delay the desired action for a few days. But in Ahaa's view, such observance was far from shared by everybody living in the area:

> We can see in our calendar when we can carry out certain acts that are usually bad [*muu*], such as forcing things into the ground, for example the *zel* [the wooden pole that holds the tethering

> line for yak calves and mares]. If we don't pay attention to this, gazryn ezed will become angry. For example, ninjas dig many holes but they don't care about gazryn ezed.

This statement left his wife speechless and visibly taken aback in disagreement. But Ahaa continued:

> They don't look at the calendar every day. They don't stop working when it is an inauspicious day for digging. So, if ninjas come by here, having earlier ignored the taboos and upset gazryn ezed, and we serve them vodka or the like, gazryn ezed may get upset at us instead of at them. That is because the ninjas are bringing their bad acts to our ger—even if we don't know what they've been doing before coming here. Also, wood thieves [*modny hulgaich*] may first go to the forest, cut down fresh trees and then on their way back stop at somebody's ger. Gazryn ezed may then get upset not at the actual wood thieves but at the ail where they stopped.

Given the close proximity to the large mining camps of Ölt and Shar Suvag, where thousands of people on a daily basis transgress fundamental taboos related to the land, herders like Ahaa constantly fear for calamities instigated by these actions. In the excerpt above, Ahaa describes how angry spirits can come to his ger regardless of his own involvement in transgressive acts. Merely by letting people into his ger and hosting them, he becomes an accomplice in the watchful eyes of spirits. Although herders are quick to blame ninjas for their own personal tribulations, they also entertain the possibility that the transgressor may be a local herder. Cases of misfortune therefore generally give rise to much speculation yet only rarely lead to actual direct accusation. The difficulty in identifying a single perpetrator is due not only to the "transferability" of spirits' anger as described above but also to the infinite number of possible perpetrators. A local saying encapsulates this: "Yos medehgüi hünd yor haldahgüi."

This saying can be translated as "a person who does not know the traditions will not be punished by the warnings." Although the saying might suggest that people ignorant of taboos do not upset spirits, people explained to me that only those unaware of the traditions would not be punished by spirits. It is not only people who are aware of local taboos that upset spirits but indeed *any* person passing through the area. As people knowingly or unknowingly upset local spirits, it becomes impossible for victims of their anger to identify its likely source. Angry spirits abound, as do possible transgressors of taboos. Thus the proverb can be seen to both expand the number of possible transgressors and facilitate the common practice among herders of blaming ninjas for their personal misfortunes by casting them as those who do not know the traditions.

Spirits and *Ails*

To this point spirit beings may seem highly oppressive and fearful, but they are also generous and cooperative. Central to this is the notion of *hishig*—that is, "blessings," "fortunes," or "favors" that spirits are seen to bestow on those who show them appropriate respect. According to Rebecca Empson (2007, 114–15), hishig refers to "a life-force or animating essence that can be understood through actions that involve attending to a part or portion that fuels a whole" (see also Empson 2011, 74–76). Since it is particularly within household clusters that hishig is nurtured and protected, I will here show how it underlines the ideal of peaceful conviviality within stable kin groups as described in chapter 1.

Herders use the word *hishig* when talking about the different rewards spirits offer people for their observance of taboos. Rather than being part of an explicit bilateral exchange, hishig is "gathered, harnessed or beckoned" through people's everyday practices (Empson 2011, 76). At times the speaker may denote a particular part of the landscape as offering hishig, such as *delhiin hangain hishig* (blessing of the world mountains). However, most often it is simply referred to as a generalized entity with no identified attachment to specific features of the landscape. My host sister Baajiimaa described hishig in the following way:

> There is hishig everywhere. In the forest, for example, our hishig is collectible wood, bushes full of delicious berries and pine trees with plenty of nuts. On the steppe [*belcheer*] it is big fat marmots, green and tall grass. In the river it is plenty of clean water. In this way, the love of lus is all the good things that make our lives possible as opposed to all the negative repercussions when we don't show due respect to lus. At such times, hishig is arguing [*hishig heleldej baina*], and we won't be able to find any berries or the like.

Among herders, hishig includes all the aspects that generate the wider basis for their pastoral economy. Without clean water or benevolent spirits, a herding household cannot sustain its herd and human members. The seriousness with which herders approach the taboos described in the previous section emphasizes this reliance on spirits' presence and generosity. Herders often express immense gratitude if spirits approve of how they go about their lives. After a successful calving season, Yagaanövgön exclaimed gleefully, "If Tsagaan Övgön [White Old Man] wasn't happy with how we live our lives, he wouldn't give us so many baby animals. He is obviously very pleased!"[5] Apart from observing taboos, respect is also shown by placing small offerings (*tahil*) on the countless stone cairns (*ovoo*) that adorn the hills and mountains of Uyanga. People offer dairy products such as *aruul* (dried milk curd; see figure 8 in chapter 2) or cooked food such as homemade cookies (*boov*) or fried flat bread (*bin*). One day my host sister and I walked up a steep mountainside with the yaks and were on our way to a watering hole located on the other side of the mountain pass. We took a rest by a small ovoo, broke off pieces of homemade cookies, and threw them on the cairn when she remarked, "Remember, what you give *has* to be homemade food [*geriin hool*]." I asked why this was so important and she said, "Otherwise gazryn ezed might get confused and not know where the offering came from. When you give offerings to gazryn ezed, they will become happy [*bayasah*] and not forget [*sanahgüi*] the people who live on their land [*gazar*]."

This close relationship between spirit beings and the locality was also brought up by Budlam, a village lama with whom I stayed for many months. Talking about seasonal ovoo rituals (*ovoo tahilga*) held at the main stone cairn just outside the village, Budlam commented, "What is important in ovoo rituals is to make offerings [*örgöl örgöh*] to all of those [all spirit beings] who are in that area. Ovoo rituals are about the area [or the land] rather than just the *shar lus* [yellow lus]."[6]

The term *hishig* is also used among local hunters when describing the process of hunting and the game they kill. However, certain animals, such as young deer (*buga*), should never be killed, and if a hunter does kill it, that prey is not described in terms of hishig. A hunter explained to me that when people do kill a young deer, they would have tricked and stolen it from *hangain lus* (the mountain lus) (see also Nadasdy 2007).[7] Such theft is dangerous because not only will spirits become very upset and increasingly hesitant to part with animals, but they will also withhold the ail's future wealth, ensuring that household members for generations to come will not become rich. Given the notion of baigal' with its interrelated beings, people cannot autonomously insist on spirits' hishig. If they try, their punishment will be severe, as the following excerpt from my field notes shows:

> Earlier in the day a middle-aged man, already drunk, came to our ger. I only made an appearance and instead sat chatting with Mum outside until the visitor eventually left. The man was called Mendjargal and I later learned that my family strongly dislikes him. He is not only a heavy-drinking herder who is seasonally involved in ninja mining but also a hunter who has killed wolves, marmots, and even deer. Mum said, "He killed *hangain delhiin hairtai yum* [the love of the world forest]. This is very bad. He killed deer, squirrel [*herem*], marmot [*torog*], white hare [*chandaga*], snow cock [*hoilog shuvuu*], and great bustard [*todog shuvuu*]. Hangain delhiin hairtai yum also includes trees, such as pine trees [*hush*], and Mend once cut down a pine tree—the whole tree! He did *üiliin ürgüi yum* [bad karma actions][8] and we say that this affects not only the person himself but several generations after him, at least three. So if someone does bad karma actions, later generations will have a poorer life. One of his younger siblings is retarded and everybody knows why that is. People here talk about how Mend once cornered a group of wild deer and as he shot each of them, even a little calf, a flood of blood [*tsusny gol*] ran down the mountain. Horrible! Because of all these bad karma actions Mend today is a *bayajihgüi hün* [someone who cannot become rich]. At one point he had many animals, but now most of them have died. This is not just him being a bad herder but one who has upset and shown no respect for lus. It doesn't matter what he tries to do. For him there is no hishig [*tüünd hishig baihgüi*]. He does some ninja mining to make money, and some hunting and herding, but he will never become rich. We say, "Having killed three deer the impoverished man will remain forever destitute" [*gurvan buga alaad bayajaagüi hün hezee ch bayajihgüi*]. The deer is such a beautiful animal. One should never kill it.

Having repeatedly ignored taboos and even killed the most cherished of animals, Mendjargel is withdrawn from the social pursuits that lead to prosperity and reduced to punishment for his own wrongdoings. Although certain archival manuscripts mention tenger as deciding the destiny of people (e.g., Humphrey and Onon 1996, 197), my host mother, Yaanjilham, who narrated the above, did not refer to this entity. Instead, in her narrative, Mendjargal and his family suffered punishment inflicted by a generalized, nonidentified being. Consequently, he became a cautionary tale about the dangers of not taking spirits seriously and a reminder to act respectfully in order to be able to attract their blessings.

In addition to pastoralism and hunting, hishig is also discussed in the context of Mongolian kinship. Empson (2003, 135) describes hishig as a life force of the ail, which all members, including daughters-in-law, serve to protect. Within the household group, hishig is evidenced in the health of the household members and their animals, as well as in general luck and fortune. However, in the case of adversities affecting the ail, daughters-in-law are often blamed for not nurturing hishig well enough. Although it is impossible to confine or contain hishig, it is generally regarded as an entity that may diminish if household members are not careful in their interactions with outsiders (Chabros 1992, 155; Empson 2012). In order to protect hishig, male household members cut hair from animals' tails before they are sold and then tie the hair to the central rope of the smoke hole of the ger. Also, before sheep are slaughtered, men rub the animals' noses in order to capture their final breath. They then rub their hands off in their deels, which are said to absorb that breath. For women, the top layer of fresh milk, *airag* (fermented mare's milk), and new tea must be kept within the household group and never given to outsiders. These practices of nurturing hishig underscore the importance of withholding the most potent part of the household's products and keeping it among the members of the ail. Just as the first bottle of home-brewed yak vodka is the strongest, so the top layer of milk (*deej*) is the fattest. The deej is seen to carry the strength of the ail, both literally and cosmologically, and if it is given away, the household group will eventually wither. Similarly, the hairs of animals displayed on the central rope of the ger enable people to marvel at the pastoral success of the household group whose deels have become thick with animal grime from the many animals that have been born, herded, and eventually slaughtered. Such daily testaments to the achievements and endurance of the ail in the face of constant negotiations with spiritual landscapes accentuate the importance of the ail as the primary social and cosmological unit.

During my fieldwork, I often heard household heads and their wives complain that their daughters-in-law were not taking enough care of hishig. In the ail of Yagaanövgön, a fierce argument erupted one day when Ahaa's wife, supposedly by accident, gave her own mother ten liters of fresh milk without first taking off the deej for the ail.[9] The daughter-in-law had a relatively small herd of milking yaks that she milked every day. The milk from her yaks sustained her husband, her son, and herself, and they decided how much milk to sell and how much to prepare for their own consumption. Living in their own ger, they thus had their own milk economy divorced from that of the rest of the ail. However, despite the separate production and consumption of the milk from her yaks, members of the household group still considered the milk cosmologically related to the ail. When Yagaanövgön found out what she had done, he shouted furiously at her, and Yaanjilham pulled her aside, threatening her against continuing disloyal acts. She and her husband apologized incessantly. But to this day, Yagaanövgön and Yaanjilham are not convinced that she is doing her utmost to nurture hishig.

These concerns about hishig highlight the ail, rather than gers or individual family members, as the primary unit involved in and affected by spirit interaction. Krystyna Chabros (1992, 155) makes this point forcefully in her study of ritual practices among different Mongol groups, based on historical manuscripts from the nineteenth and twentieth centuries as well as ethnographic fieldwork. She states that "*Kesig* [*hishig*] represents the individual's share (portion) of the vital energy of his lineage. The shift in meaning in the word has been from a concrete sense, in which a portion of a larger quantity of meat represents the relationship between a man and his clan, as part to whole, to an abstract sense in which the word refers to the energy which animates the clan and of which each individual member partakes."

Although in-laws can never become agnatic kinsmen, they are expected and pressured to act as such in their care for hishig. By acting according to the interests of kinsmen, the in-marrying daughters-in-law protect hishig as well as the future descendants of the ail. Kin ideology is in effect supported in this way by spirit reckonings, emphasizing the ail and its necessary affines as an ideally cooperating and harmonious unit. But at the same time, a rebellious daughter-in-law can distribute or neglect hishig, thereby affecting the ail in the most fundamental of ways by preventing the flow of its future fortune.

Hishig and Gold

In Ölt and Shar Suvag, ninjas also talk about and express concern for hishig. However, their meaning for the term is slightly different from that used by herders and local hunters. One day, when I was talking with a group of ninjas, one of them known as Oyuun said, "We never find more than just enough gold to pay for our daily meals and the like. The hishig of our district doesn't seem to be able to reach far enough [*sumyn hishig hürtej chadahgüi yum*]." I asked her what she meant by *hishig*, and she elaborated:

> Gold! We never find any gold. This area has no gold anymore. Also trees! People come and cut down all the trees so there's no firewood left for us. And berries! People pick all the berries so we have nothing. We can't eat berries in August like we used to. We can't make berry wine for Tsagaan Sar [the lunar new year] anymore. There's nothing left for us. And nuts! People come out here and fill so many bags with nuts that they take them away on jeeps and trucks. They then sell the nuts in the cities and make a lot of money, while we have nothing to eat. There is no longer any hishig to be found!

In referring to gold as hishig, Oyuun underscored that the precious metal, just like berries and trees, is part of the physical environment and forms the basis of human livelihood. Emerging from the landscape, gold is considered an object of fortune that is neither predictable nor controllable. The mines has its masters who decide (*shideh*) or administer (*udirdah*) how much hishig people are likely to attract, and as a result, finding gold is not simply a practical matter of digging deep holes and panning the unearthed gravel for gold. The amount of gold ninjas find is instead perceived as necessarily dependent on the generosity of spirits. Knowledge of mining techniques and local geology certainly enhances the chances of hitting a gold deposit, but such insights alone are not deemed sufficient. Maintaining a good relationship with local spirits is thus paramount to the mining successes of ninjas. However, many ninjas recognize that local spirits do not approve of their mining activities. They reason that because of this disapproval, spirits withhold hishig, thus making it harder for people to find gold. Furthermore, as the ninja population increases year after year, the pressure to attract the already diminishing hishig grows correspondingly, and some ninjas end up searching for gold in vain.

The notion of protecting and nurturing hishig, which is so central to Uyanga's pastoral economy, does not seem prevalent in the mines. During my fieldwork I did not come across practices that served to preserve or cultivate a smaller "portion that fuels a whole," as Empson (2007, 115) described the nature of hishig. For example, there was no sense in which the first gold extracted from a mining hole was considered a different kind of affective substance from the remaining gold, that is, as some kind of "gold deej." Nor was there a view that a portion of gold should be held back in some way, just like the hair from animals' tails. On the contrary, all the excavated gold is sold within the day to gold traders, and ninjas are reluctant to store any gold overnight in their gers (see chapter 4). Apart from increasing the risk of theft, storing the gold exposes them to the so-called power of gold—that is, the invisible dust (*haradaggüi shoroo*) that rubs off gold during excavation, which I discussed in the previous chapter—and this is not a desirable residue akin to animals' grime and final breath. Rather, the residual dust is feared for its ability to pull ninjas against their own will, rendering them weak and powerless. Although ninjas talk about gold in terms of hishig, it is thus a fundamentally different kind of substance from those found in pastoral and hunting economies.

In the mines, spirits not only withhold hishig but also wreak havoc by inflicting illnesses and causing accidents and even deaths. When a ninja experiences such misfortune, it is usually said that it is because of the anger of lus. Given the constant presence of angry spirits in the mines, rumors circulate that at least one ninja dies every single day. Fortunately I did not come across any actual deaths, but in a situation in which people know that dangers are lurking everywhere, they talk a lot about fatalities and suffering. I was warned when passing deep mining holes (figure 12) that a corpse might lie rotting at the bottom. Such corpses are considered particularly dangerous as they turn into "evil souls" (*chötgör*), hungry for the human life in which they once partook. Since ninjas repeatedly transgress taboos, they are seen to accumulate bad karma actions. When they die, it is thus particularly important to dispose of their corpses properly and to carry out, to perfection, elaborate funerary rituals. In case of even the smallest oversight, the deceased is said to turn into an evil soul, haunting the living. In order to ensure that a potential chötgör will not trouble the residents of Uyanga, local lamas place corpses outside Uyanga in a neighboring district (*sum*). In return, the lamas of that district can dispose of their corpses within Uyanga. In this sense, potential evil souls are exchanged between neighboring districts. Given the unmatched concentration of ninjas in Uyanga, locals often laugh at this "exchange" since it works out to the definite advantage of Uyanga residents.

Living in such a seemingly doomed environment, ninjas often request that lamas carry out ritual offerings and recitations of mantras that serve to beg and cajole (*guij argadah*) spirits into bestowing hishig. During my stays with Budlam, who was one of the senior local lamas (see chapter 5), I observed that ninjas constituted by far the largest majority of people who requested his help. With ninjas having become his most frequent and loyal clients, his frustrations with their way of life seemed to be growing. It was my impression that these frustrations were shared by many of the other lamas at the village monastery. One day, when the wave of clients requesting

Figure 12. A mining hole in Ölt

that he do "personal readings"[10] finally seemed over, he leaned back on the chair and sighed:

> Oh, so many of these ninjas have a wrong understanding of life and religion. They come here to ask for hishig or they ask me to go to the mining area. Maybe the ninjas are starting to find less gold or are falling ill. But they always seem to think that by giving offerings [*örgöl örgöh*] and by asking for *lus savdag*'s protection they will find more gold. This is wrong. They don't understand that it is their own greediness [*shunal*] that causes their misery [*zovlon*]. Gold is very dangerous; they should leave it alone. They don't need it because it always leads to problems. Gold is not like silver or other metals. Gold is heavy and potent [*hüchtei*]. People can't protect themselves against the power of gold and it ends up making them see nothing but gold. All they think about is gold. All their lives are about is money. To give offerings doesn't make the lus savdag blind to their disrespect for nature. They have a wrong understanding of the teachings. Greediness is one of the principal paths to misery. If one studies the teachings, one will learn that by caring about everybody, you yourself will also be cared for. All of us can become

> Buddhas, we all have the potential. So by not thinking so much about yourself, because that's exactly what greediness is, right, you will be protected [*hamgalagdah*] by the Buddha, whose teachings tell people to show compassion [*nigüülsengüi*] and wisdom to the benefit of all.

Budlam's statement not only illustrates the extent to which local spirits can be approached and related to in multiple ways. It also defines lamas as religious specialists positioned centrally between spirit beings and ninjas. Local lamas complain about the intense attachment of ninjas to a materialist world in which they willingly pursue momentary prosperity at the cost of upsetting spirits and fellow humans. Yet alongside their criticism of ninjas, the village lamas can essentially be seen to facilitate gold mining because their help is deemed necessary to prevent the wrathful spirits from becoming fatal obstacles to ninjas' pursuits. Without the safeguarding of lamas, ninjas would probably soon give up their search for gold.

The Black *Lus*

When I first arrived in Ölt and Shar Suvag, I was immediately made aware that spirits roaming the mines were quite different from those of the surrounding areas and that I should adapt my behavior accordingly. In the mines, a row of rocky crags locally known as Hairhan (lit. merciful or gracious, a commonly used term to refer to sacred mountains) overlooks the mining camps. When twilight approaches, groups of male and female ninjas of all ages climb the steep path to the rocks, where in the dark evenings endless lights can be seen from afar. One day my host mother, Nergüi, invited me along on her visit to the offering place. It was early evening and starting to get dark. Accompanied by some of her friends and children, we walked up the steep hill to Hairhan. When we reached the plateau just before the last climb, we encountered a large family already placing offerings at the rock. They lit candles and took turns praying in front of a bundle of blue ceremonial silk scarves [*hadag*] that were tied around a small part of the outcrop. With their palms pressed together, they touched the rock three times with their foreheads and mumbled something inaudible. The atmosphere was solemn. They did not speak among themselves; even the young children were completely quiet. The father carried a bottle of vodka and the daughter a bottle of *undaa* (soft drink). When they had lit all their candles, they descended from the offering place and climbed clockwise around the outcrop. They stopped briefly at a small ovoo at the top of the rock before continuing down the other side. Another group of people were sitting some distance away from the rock. Other people sat farther down, having not yet climbed up to visit the rock. Everybody sat quietly in the darkness, waiting for people to leave. Group by group, they took turns visiting the offering place, and eventually it was our turn. We climbed up and visited the rock.

Visits to Hairhan constitute a significant part of ninja ritual and social life, involving nuclear family members, neighbors, and other friends. On the walk from the loud and bustling mining camps to Hairhan, there is much banter and laughter, but as the crags begin to tower high above the path, voices lower and some of my hosts often begin to share their thoughts. Tsegii from my mining team once reflected, "Bataa hurt his leg badly yesterday. He was at the *pajur* (rocker, a type of sluice box), when a sharp stone hit his leg. Lus is angry." Another had had unrelenting abdominal pains and hoped that offerings to lus could bring about some much-wanted relief. Noting that ninjas often make offerings for very specific reasons, I asked my friends if they sometimes begged (*guih*) lus for a more general sense of fortune and cosmological benevolence. Most of them simply replied no (*ügüi*), but some intimated that ninjas did not have that kind of connection (*holboo*) with lus. A ninja from Darhan thought that if it happened, it was probably only ninjas from Uyanga who could make such requests. Since people often stay in Ölt and Shar Suvag for only the warmer summer months, I suspect that ninjas are unlikely to have established the long-term bilateral connections with spirits that are so central to human-spirit relations in the region. Moreover, during their stay in the "land of dust" ninjas are constantly working their way toward new lands. There is little sense in returning to an already-worked mine when gold-bearing deposits lie ahead. In contrast to herders, hunters, and villagers, ninjas thus stay in an area only for as long as necessary. When I first noticed that ninjas made offerings of vodka, I wondered if maybe this transient nature of mining life was part of the motivation for these stronger offerings to spirits. But during an extended drinking binge, my host father Batzaya opined that "*har lus* is just like us. Of course [it] likes vodka more than milk tea!" before we burst into laughter. The fortune-teller Dorsüren, who usually offered lengthy talks about all kinds of matters, also had little to say about this. One day when we talked about the fact that dairy products were not used for offerings in the mines, she told me the following: "Oh, you mustn't sprinkle dairy products [*tsagaan idee*] here. Only vodka. You know, there are black and white lus, and if people don't show enough respect for them, the good white lus will leave. So when people dig deep holes and destroy the river, the white lus leaves and only the powerful [*hüchtei*] black *lus* remains."

"The black *lus* . . . ?" I asked.

She continued: "Black lus is the one that can make people fall ill and cause other bad things to happen. To appease black lus you must give vodka; milk or tea is for white lus. That's also why you must only give offerings at night. Never during the day."

These particular ritual practices highlight the marginalization of largely benevolent white lus and the emerging dominance of the black lus. In the mining and herding areas, it is generally agreed that this black lus did not previously exist in the area and arrived only with the onset of the gold rush. Although black lus is indeed mentioned in Tibetan Bon manuscripts (Hildegard Diemberger, personal communication) and Mongolian religious texts (Bawden 1994, 70), it is still described locally as new (*shine*) and dangerous (*ayultai*).[11] As with the emergence of any spirit entity, black lus has opened up a space within which unknown powers and possibilities have appeared. In their recognition and veneration of black lus, ninjas have entered into a relationship with and exposed themselves to these forces—forces that are still relatively little known among the people of Uyanga.

In the previous chapters I have argued that the gold rush has offered an opportunity for local herders to achieve distance from their extended kin groups. By joining the search for gold, they have become part of mining teams that allow for, and in part generate, an unprecedented degree of individual autonomy and independence. Important displays of this social position of ninjas include their acquisition of new technical skills, mineralogical insights, and specialized vocabulary. But it is also grounded in their rumored ability to engage in new forms of "black magic" (*har dom*).

As ninjas climb up to Hairhan and worship black *lus*, they are held to come into contact with an intensified form of a highly potent substance known as *altny gai*—the "misfortune of gold."[12] This substance is said to flourish in the mines and can allegedly be redirected onto others through ritual action, causing illnesses among its victims. There is much fear of such attacks, and people rarely discuss specific instances of malevolent ritual practices carried out by ninjas. As a result, the only specific situation I heard about involved me as the victim. The situation I describe below concerns my initial host family on the steppe, whose members were either ninjas or in other ways involved in gold mining. As I later learned, my host family had a reputation for being dangerous because altny gai was said to flourish in their ail.

It was the second winter of my fieldwork, and temperatures were finally starting to rise above minus thirty degrees Celsius. I was staying in Yagaanövgön's ail and had pleaded with him for permission to revisit Nyambuu's (my first host family in Uyanga). Finally, my host father gave his consent. Ahaa took me by motorbike across three mountain passes to the other family, where we were welcomed with plenty of hot salty milk tea and seats by the warm stove. As Ahaa took his farewells, we agreed that he would return one month later to pick me up.

During the following week I helped my former host mother, Degidsüren, with the usual household chores and animal duties. As we worked together, she asked me numerous questions about Yagaanövgön's ail. How many animals did they have? How many motorbikes? Were any of their children involved in ninja mining? Following local custom, my replies were minor lies, slightly distorting reality in order to avoid having my answers used in malicious gossip. Degidsüren repeatedly asked me what daily life in their ail was like and how they treated me. As I admitted that they treated me well, she immediately replied that they just wanted my money and that they were stingy (*nariin hün*) and selfish (*aminch*). She brought up the subject several times, each time as a long, recurring monologue. A few days into my stay I suddenly fell ill with high fever, and as the days passed I felt increasingly worse. Since my health had generally been good and I could not identify any symptoms other than the fever, I grew concerned. All members of the family said my illness was the result of my low blood pressure (a common explanation in the area) and that I should just eat extra fatty food. After a few days on a strict diet of boiled mutton fat, I surrendered and said that I could no longer cope with the food. I had had enough boiled mutton fat and had to get back to my host family, where my medical kit was. I needed medicine. The family was reluctant to let me go on the long and cold motorbike ride in such a feverish state, but they eventually gave in.

When I arrived at Yagaanövgön's ail, my host sisters pulled me inside quickly and warmed me by the stove before laying me on the bed and showering me with questions. I was exhausted, delirious, and tired. Yagaanövgön appeared, drunk. He sat by my side and, after asking what was wrong, he raged: "Nyambuu's have gone mad [*Nyambuuginhan galzuurj bolson*]! They have cast curses on you [*ted chamd har haraal hiisen*]. That's why you are sick!" I grew increasingly uncomfortable and uneasy—why would they cast curses on me? How would I get healthy again? As I sweated through my fever the next couple of days, the questions multiplied and seemed to grow in urgency. After I recovered, my family started telling me more about my illness, which had been combated with a mix of my antibiotics and their incense, mantras, and offerings. They told me that since most of Nyambuu's children work in the mines and often come back with ninja friends, they bring with them trails of altny gai. At Nyambuu's there is thus altny gai everywhere, and it can be redirected at people, attacking them until they fall ill and even die. Yagaanövgön asked everybody else to quiet down so he could explain: "You have stayed with us for a long time now. You are not from their ail anymore. You have become vulnerable, like a child. If Degidsüren wanted to, she could make all these invisible beings [*haragdaggüi yum*] and evil souls [*chötgör*] attack you."

I asked why she would ever do something like that. "Oh, she is a very jealous person," an explanation with which they all agreed. The incident became a shared secret within my host family and was brought up only when a family member expressed interest in going to the mines.

The secrecy surrounding such malevolent attacks is in no way unique to the Mongolian cultural region. Indeed, the anthropological literature on witchcraft and shamanism in highly disparate areas describes the common reluctance among informants to openly discuss these matters. This reluctance may be grounded in a fear that listeners will presume that the knowledge an informant imparts reveals his or her own possible involvement in such matters. In the Mongolian context, such an association between knowledge and personal practice may be particularly pertinent since, as Humphrey and Onon (1996, 324) notes, "To know [is] to understand and have power over the object known. The Mongol and Daur verb *mede-* means both to know, recognise, or discover and to manage or rule."

In this sense, knowledge is not simply a question of insights but also capabilities. "To know" is an active position of personal agency, which reveals much more than the casual accumulation of information. It is thus not surprising that ninjas rarely talk about their reputed capabilities to redirect the misfortune of gold. The extent to which their involvement with altny gai is desired, and even possible, is therefore hard to discern.

However, regardless of their actual skills in malevolent ritual practices, ninjas are greatly feared for their power to inflict illness and even death onto potential victims. The invisible beings that ninjas are considered capable of addressing are not the largely benevolent spirits of the steppe. Ninjas do not engage gazryn ezed, usan khan, Tsagaan Övgön, or tenger. Instead, they primarily interact with har lus, if not the chötgör that congregate around the deep mining holes with rotting corpses. These invisible beings seem to accumulate their potency from precisely those areas where ideas of communal living are negated. These entities do not reside in mountains, rivers, or trees—that is, identifiable markers of the landscape. Instead, they thrive in what are locally called "areas of flow" (*güideltei gazar*) or "areas with no master" (*ezengüi gazar*)—expressions that are also used for burial grounds in both rural and urban areas (Grégory Delaplace and Christopher Kaplonski, personal communication). By addressing and amassing all the uncontrollable and invisible beings that roam this landscape, ninjas are feared for their powers, which surpass those of other people. Neither lamas nor older patriarchs claim such power. Rotting corpses and distinct ritual practices thus convey not only an altered interaction with the landscape and its taboos but also an engagement with different kinds of spirit beings. By challenging the expected positions of respect and humility vis-à-vis the forces of baigal', ninjas thus position themselves beyond the hierarchies of both kinsmen and local spirits.

In this chapter economic life emerges as a manifestation of human and nonhuman interaction. It is only by nurturing relations with spirit beings that the desirable gold or the fresh berries will reach people. By pivoting on the observance of taboos, the placing of offerings, and the consultation with calendars and religious specialists, economic life is a moral practice that is not only relational but also transhuman. But this is not to say that people in Uyanga are particularly moral or that people's morals should be defined solely by the way in which they make a living. As demonstrated in this chapter, peaceful relations between beings are a challenge, not a condition, and for some the challenge is insurmountable. Indeed, with powerful black spirits having arrived in the mines, alongside possibilities for new forms of malevolent ritual practices, the mining boom offers not only unmatched opportunities for wealth creation but also a constant risk of misfortune. I have shown that in a cosmoeconomy in which the tense symbiosis between humans and nonhumans is not grounded in consensus or harmony, the desire for the precious metal demands recognition of powerful spirit worlds.

These cosmological dangers not only concern illnesses, accidents, and deaths but are also transferred onto the money earned from gold mining so that dirty, crumbled money notes become an index of potential pollution. The following chapter thus shows how money, although posited as an instrument of value equivalence, is intimately tied to local moral understandings of gold mining.

4 Polluted Money

IT WAS A summer's evening, and I had been selling fresh yak milk all day in the mining camps with Ahaa, the oldest son of Yagaanövgön. We had left their *ail* in the early morning with several large milk containers strapped to the sides of his motorbike and returned only once we had sold all the milk. We took a seat near the warm stove and were each given a cup of soothing milk tea. Ahaa pulled out a sizable bundle of soiled money notes and handed it over to his father. Yagaanövgön silently glanced at the notes and tried to straighten them out and brush off any caked mud. He sighed deeply, then looked up and said sternly, "This is the last time we sell milk in the mines. From now on we will only sell it in the village. Do you understand? No more gold money [*altny möngö*]."

Ahaa began to protest, remarking on the higher price they receive for their milk in the mines. But Yagaanövgön continued, "Don't you understand? Haven't you seen enough? Your son has been sick, your sister has been sick, your mother is sick. Why? Why do you think this is so? Stop bringing this polluted money [*buzartai möngö*] into my household. It's pure rubbish! I've had enough! Enough!"

This view of money earned from mining as in some way dangerous and inauspicious is not unique to the Mongolian gold rush. Indeed, a closer inspection of ethnographic descriptions of mining carried out elsewhere in the world reveals a recurring moral tension between the extraction of minerals and the monetary wealth it produces. In gold mining towns in Kenya people are concerned about the circulation of "bitter money" (Shipton 1989, 37); in Angola diamond hunters seek to tame "wild money" (De Boeck 1999, 187); in Papua New Guinea gold rush miners handle "wasted money" (Clark 1993, 744); while in Madagascar sapphire miners produce "hot money" (Walsh 2003, 299; see also Znoj 1998 201–4).[1] Extracting and selling minerals often give rise to a distinct form of monetary wealth, a wealth that has elaborate, but not necessarily uniform, symbolic connotations that can affect the kinds of exchange relations in which it will later take part. Whereas economists consider that state currencies form a "frictionless surface to history" (Graeber 1996, 6), polluted money in its various forms retains strong attachments to its origins. With its multiple ties to the past and present, it circulates as a powerful moral compass of human actions. Appearing alongside cosmological evocations and angered spirit beings, polluted money underscores the moral strains that people often face when maneuvering within cosmoeconomies of mineral extraction.

Drawing on Alaina Lemon's (1998) view of wealth objects as "affective and aesthetic matter," in this chapter I examine pastoral wealth and polluted money less as abstractions of exchange value than as objects that retain strong attachments to the spirit worlds described in the previous chapter. Rather than presuming that these wealth objects are categorically dissimilar, I show how they both figure centrally in people's negotiations of envy (*ataa*). The wealth that people associate with pastoralism is largely visible, countable, and comparable. This material form affects not only how it is perceived and handled but also how vulnerable people consider themselves to be to attacks of *hel am*—a form of "speech acts that inflict various degrees of harm" (Swancutt 2012, 127), which Lars Højer (2004, 81) refers to as "the witchcraft of ordinary life." However, the money that people earn from gold mining does not offer a simple solution to their predicament. Circulating as a feared currency, polluted money is deemed a carrier of misfortune that has now brought into question commonplace calculations of monetary equivalence. This is not just another example of money's being symbolically earmarked. It is a differentiation that also permeates the fiscal properties of money and affects daily transactions. In the village shops, cash value has thus become a fluid and unstable category that is both informed by money's formal position as legal tender and intimately tied to local moral understandings of gold mining. Although posited as an instrument of equivalence, the money that now circulates through Uyanga is far from being uniform.

Pastoral Wealth

The herding households of Uyanga tend to have yaks, sheep, horses, and varying numbers of goats. In the early 1990s, many herders were so excited about becoming so-called *huviin malchin* (lit. private herders) that they also purchased camels and cattle. Nyambuu, my first host father on the steppe, reminisced that "we could suddenly have any animal we wanted!" Despite the euphoric attempts to introduce camels and cattle into this high-altitude region, unfortunately not a single one of them survived. It has now been more than twenty years since the herding collectives (*negdel*) of the socialist period were dismantled, and people today rarely talk about current herds as in any way related to the past division of herding-collective livestock. Instead, people see large herds as reflecting admirable herding skills, peaceful relations with spirit beings, and the good general conduct of an ail. A herder can thus be highly skilled and have detailed knowledge of pastures and seasonal changes but still not manage to sustain his herd. Herders are generally reluctant to disclose to visiting relatives and friends the size of their herd (see High 2008b). For Yagaanövgön and Nyambuu, it is a highly sensitive, if not secretive, matter that reveals personal dynamics involving households and spirit worlds. As the previous chapter illustrated, the ability to attract and nurture spirits' *hishig* is related to the peaceful conviviality within stable kin groups and their maintenance of respectful relations with powerful spirit beings. Whenever they are asked the delicate question about herd sizes, Nyambuu and Yagaanövgön thus invariably pay no heed to the question and shrug it off by simply replying, "Oh, Tsagaan Övgön has been good to us." They then often try to turn the conversation on to others in the area. In their discussions, they rank them: the first household group (*negdügeer ail*), the second household group, the third, etc., according to estimated wealth. The order is under casual discussion in visiting situations, as the following excerpt from my field notes shows:

> In between numerous drinks and cigarettes, a close friend who paid daily visits to our ail said to my host father [Yagaanövgön], "I know you think he's the fourth, but have you seen the large winter shed he just built? His herd must be much bigger than you think. I'm sure he's the third now." A dry laughter filled the air, a cigarette butt was thrown in the direction of the stove, and another sip of vodka was downed before my host father replied, "Well, you'll see. I pass his herd every day as I take my sheep out on the pasture. I trust my own eyes more than your fast tongue!"

These comparisons of wealth concern not only the number of animals but also other household possessions such as animal sheds, motorbikes, solar panels, and TV satellite dishes. Attention is fixed on the substantial wealth that is there for everybody to see. This visibility of pastoral wealth became apparent to me one day when I was riding through a distant valley with Yagaanövgön. With each ail we passed, he provided an elaborate inventory of their possessions. I was surprised by his in-depth familiarity with households that I did not know he even visited. "It's easy," he said and then explained the following:

> When you see an ail you can immediately tell which animals they have got, how many of each, how they use the animal products, whether they are good herders or not, how many sons and daughters they have, etc. It's all there for you to see [*harah*]. Is the area around their *gers* clean? What kind of garbage [*hog*] do they have lying around? Do you see any tracks from tires? Even if you've never visited the ail, you can still tell how wealthy they are. Here nothing is secret [*büh yum nuutsgüi*].

Herders estimate not only the material and animate possessions of ails but also their financial potential for making new purchases or investments. Long before a large purchase or investment is made, gossip crosses the mountains and reaches remote ails, enabling their members to discuss in minute detail whether such spending really is sound. Over a serving of freshly made vodka, people discuss the rumored earnings from the sale of cashmere (see figure 13) in Arvaiheer or the expected monetary return on the temporary loan of a milking yak to a neighbor. Cashmere is one of the main sources of income for herders in Uyanga, but they also receive money that is not generated from livestock, such as seasonal employment income, pensions, or other forms of social welfare support. According to Nyambuu, even such monetary flows are well known and easy to predict: "We all know who receives a pension here. We all know how much cashmere each of us sells and how much we sell it for. We all know who pays which taxes. We all know who pays health insurance. And insurance for their motorbikes. And for their cars. We even know who ends up paying a fine when they are caught by the police."

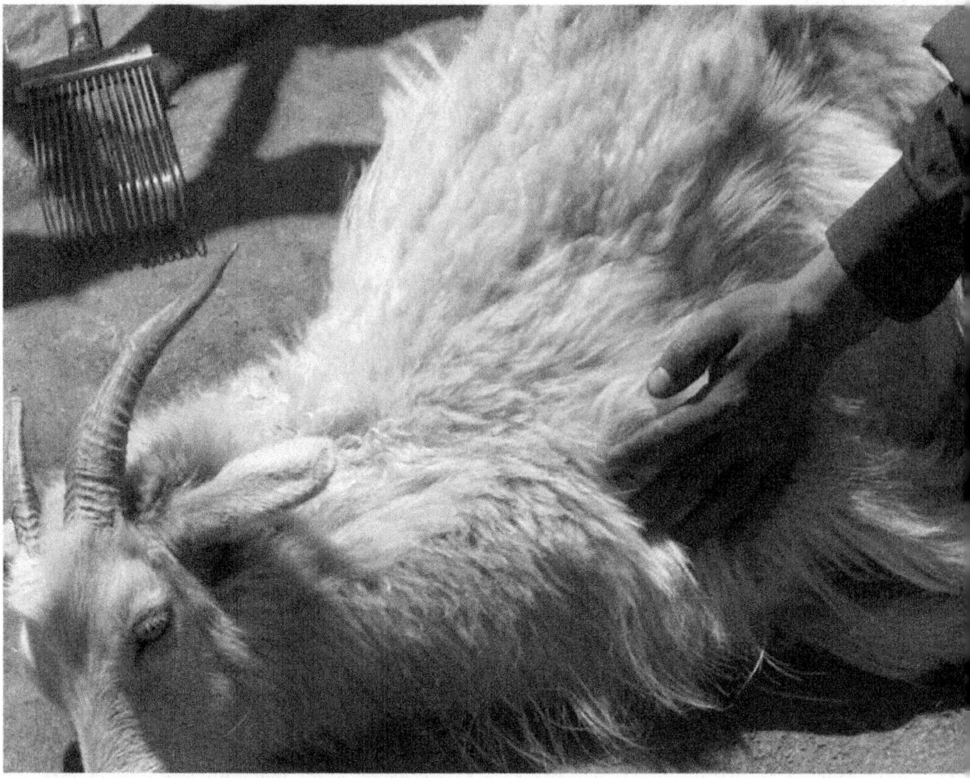

Figure 13. Shearing goats for cashmere

Although herders are reluctant to discuss their own wealth, they appear to have great knowledge about the wealth of others. Not surprisingly, this seemingly well-known and visible wealth of herding households indexes to a certain extent local expectations of generosity (*ögöömör zan*). Approached through highly respectable address and deferential demeanor, wealthy herding household heads frequently receive requests from visitors for monetary and material charity. In both Nyambuu's and Yagaanövgön's households, distant relatives and acquaintances arrive almost daily with the explicit purpose of asking for help, ranging from generously reduced prices for animals to particular material gifts. For Yagaanövgön, these requests are like testimony to his recognized accumulation of wealth, and he feels that by giving, he performs good deeds (*buyany beleg*) that can help ensure a good rebirth (*dahin törölt*).

But not all members of his ail support his desire to be generous. Since younger members can obtain a share of the household's wealth only upon marriage or inheritance, they often prefer to retain wealth within the ail rather than succumbing to the numerous requests from visitors. Moreover, in the case of daughters-in-law, although their dowry animals (*injiin mal*) are legally regarded as their own, in practice they form part of the ail's wealth and may be at risk of being given away. As a result, acts of generosity can jeopardize claims to wealth advanced by both kin and affinal household members.

As described in chapter 1, it is generally only the household head who makes decisions about monetary and material matters. When visitors advance their requests, they often give elaborate and emotional descriptions of their dire situation with so much detail that it seems impossible for the household head to dismiss their requests smoothly and easily. After such accounts, visitors will emphasize the unparalleled wealth of the host and the ease with which his generosity can be shown. The decision is thus presented as a question of the household head's individual willingness to give and not an issue of his material ability or social interest. A visitor's request presents in this sense not just a material burden but also an invitation for the household head to display his authority domestically as well as publicly. By avoiding envy and accepting the requests of visitors, he effectively positions himself at the apex of the local "topography of wealth" (Ferguson 1992).

Malicious Gossip

With visitors' expectations and demands appearing ever growing, it seems that herding households could end up surrendering all their wealth. People like Nyambuu and Yagaanövgön must therefore somehow decide the degree to which they show generosity. Yagaanövgön had a peculiar strategy for circumventing the social and moral pressure to be generous. During any given day of my fieldwork he spent most of his time away from the main ger, which guests associated with him, and instead slept, worked, and relaxed in a small, decrepit ger erected far behind it. When visitors arrived, they asked where the household head was, and the reply was always the same: "He went out herding the sheep [*honynd yavsan*]," even when everyone knew that he was sitting in the other ger. Often, once the visitor had left, Yagaanövgön would then sneak back into the main ger for a cup of milk tea and a chat. In this way he evaded visitors and in particular their demands for his generosity. In his absence it became impossible for visitors to make requests.

However, it is not everybody who has extra gers to hide in or would even be willing to do so. In these situations, when household heads refuse visitors' requests, the visitors may become envious of the ail's wealth and start spreading *hel am*, or malicious gossip.[2] I first came across hel am at the height of winter during my second year of fieldwork. Having not heard any mention of it for more than a year, I suddenly caught a quick comment by one of my host sisters. Her horse had died and she reflected, as if to herself, that it was perhaps because of hel am. I asked her what hel am was and she reluctantly whispered, "Hel am arises when people say bad things about others. If people are envious [*hor shar*, lit. yellow poison] or jealous [*jötööch*] of somebody, they might for example wish them misfortune and cause either specific things like sickness or death onto the ail or general misfortune. It is most often rich [*bayan*] people who are targeted with hel am."

As in many other parts of the world, malicious gossip is not something people talk about openly. As I grew closer with my host families, much of our confidential conversation came to concern fears and rumors of personal hel am attacks. In line with the spirit anger discussed in the previous chapter, these attacks are not so much temporally bounded events as delayed, drawn-out possibilities of harm that can happen at any time. It was explained to me that any person, regardless of age or gender, can instigate malicious gossip. What is needed is in-depth information about the target, detailing his or her family situation, extent of kinship ties, number and kinds of animals, work chores, and material possessions. According to one of the village lamas, the required information depends to some extent on the perpetrator's motivations for spreading the malicious gossip. Attacking someone because of a broken business deal, for example, requires slightly different information than a long-standing family dispute. Disclosing that much of Uyanga's hel am is related to envy of the target's wealth, the lama explained that in these cases perpetrators have to know comprehensive details about monetary, material, and animate wealth, including spending levels and past investment, savings, and extended loans. The more detailed the information, the more powerful the malicious gossip is said to become. Given the reluctance among herders to reveal the extent of their wealth, these insights require frequent visiting and an extensive network of trustworthy friends and relatives. Thus it is hard to imagine that anyone other than the household head would be able to carry out potent hel am.

Hel am does not bring about calamities like droughts or extreme winters like the long *zud* described in chapter 1. Nor does it indirectly inflict human misfortune, such as through encounters with thieves or murderers. Instead, it is seen to directly attack the vitality and prosperity of living beings. In these instances, people in Uyanga readily point to hel am as a possible explanation, as exemplified in the following excerpt from my field notes:

> As we were herding the yaks, Ber started telling me about how two of her children had died prematurely. "My children were born too early but they could have survived, I'm sure . . . but there was a lot of hel am. . . . Those very years many of our animals also died and life was really difficult for us and my parents." She spoke with such urgency that there was no time to ask questions. She continued explaining that after her first child died, she was worried that something bad would happen to her second child. She therefore insisted on giving birth at the hospital in the regional capital [Arvaiheer], which is generally considered much better than the local hospital in Uyanga. They also had lamas carry out rituals in order to ensure the protection of the following newborn child against general misfortune and hel am in particular. They were very afraid something bad would happen, knowing that a lot of strong hel am was circulating about them. Yet the second child also died and the grief was unbearable. Finally, Band was born and survived longer than the other two, but still to this day the daughter-in law is really worried that hel am will strike again and take their only child away from them.[3]

In this situation the daughter-in-law expressed an unusual degree of certainty that her children had been killed by malicious gossip. As mentioned in chapter 1, her father was a ritual specialist who carried out black magic for others (*har zügiin lam*, lit. black direction lama), and her family was disliked by many. Although her father had passed away, her family still seemed to be victims of much suffering. In order to protect their family, she and her husband now make daily offerings to their "guardian spirit" (*sahius sahigch*), burn incense, recite mantras, and request village lamas to carry out protection rituals. They have also become particularly secretive about their wealth. Since they live in an area where people can immediately observe, count, and compare significant proportions of pastoral wealth, it is not surprising that Ahaa, mentioned in the chapter's introductory vignette, is so keen to sell his milk in the mines. When they convert their milk into unknown amounts of concealable money, their wealth thereby evades easy quantification, rendering hel am attacks more difficult.

Mongolian Currencies

Given the stakes involved in the politics of wealth and envy in Uyanga, the concealable and largely unpredictable monetary wealth afforded by the gold rush is welcomed by many—not least by Ahaa and his wife. But this does not mean that state currencies are associated particularly with the advent of the mining boom. A brief look back in time allows us to see the long-standing presence of and remarkable variation in the state currencies that have circulated in Mongolia.

When Chinggis Khaan united the Mongols in 1206, he initially used cloth bolts stamped with his seal as currency. These bundles had circulated for centuries in the vassal state of Turfan, which had been an important trade center on the Silk Road (Atwood 2004, 563). Shortly before his death in 1227, Chinggis Khaan authorized the introduction of paper money backed by silk reserves, and his successors soon began to establish an increasingly elaborate monetary system for the growing empire. Rather than seeking to implement a single currency and standardize all tax payments, the Mongol khans allowed their subjects to mint coins in local metals, weights, and denominations. As a result, multiple and mutually exchangeable currencies circulated throughout the large empire.

When the empire crumbled at the end of the seventeenth century, Mongolia lost both its independence and sovereign money. In its place, "bricks of tea" (*shahmal tsai*) became the dominant medium of exchange (Enhbat 2010, 4). There were different kinds of tea, distinguished by color, thickness, and shape. They were pressed into blocks of various weights, and conversion tables outlined their exchange rates to Chinese dollars and Russian rubles. The main unit weighed two kilograms and could be divided into half, quarter, or even smaller pieces. Yellow tea constituted the lower denomination, and its ratio to the main unit varied greatly over the centuries. Since the value of the currency derived directly from the commodity out of which it was made, transactions could require large amounts of tea. Transported in heavy wooden cases holding between twenty-seven and thirty-six bricks of main unit tea, the tea currency was cumbersome and often required vehicles for its transportation. By the end of the nineteenth century, Chinese and Russian trading companies as well as Mongolian

Buddhist monasteries began issuing "bonds" (*tiiz*, seal, stamp, or ticket) that were backed by tea reserves. These were passed on through circuits of exchange and operated much like banknotes in large trade transactions and debt payments. Alongside the tea currency, silver ingots, silk scarves, and sheep were also in wide circulation as money, as were foreign currencies such as Spanish and Mexican pesos, British and American dollars, German marks, and Japanese yen.

In the early 1920s, when more than twenty million foreign notes circulated in the country, Mongolia for the first time pursued strict monetary regulation and standardization.[4] Under the Soviet socialist regime, the Mongolian Bank of Trade and Industry introduced today's state currency, the *tögrög*. It was issued at a value equal to one Soviet ruble and was convertible to silver coins. However, in 1929 the government canceled its convertibility because of the lack of state funds and turned the tögrög into a fiat currency. Its value was thus no longer based on the value of a specific commodity but rather decreed by state law. Positioned as legal tender, the tögrög now derived its value from the government's declaration that it constituted the only national medium of exchange, unit of account, means of payment, and store of wealth. The tögrög followed the course of many other fiat currencies and experienced periods of high inflation (Rolnick and Weber 1997). In the early 1990s inflation soared to such an extent that all coinage was withdrawn from circulation as a result of its negligible value, and the Mongolian state currency today thus consists exclusively of notes.[5]

Throughout the history of Mongolia's many currencies, gold has never been used for the minting of coins or as the corresponding standard of value.[6] This is despite the fact that gold is historically one of the most common referents of value for currencies. In many countries gold has been used as the metal for coinage or, with the adoption of the international gold standard in the late nineteenth century and its later modifications, as the pegged standard of value. However, it is with the advent of the Mongolian gold rush that the value of the tögrög has become directly affected by the unearthing of the country's gold, albeit in ways that differ significantly from the metalist vision of value and its link to precious metal.

Buzartai Möngö

Throughout my fieldwork, Buyanaa, the younger brother of Ber, lived and worked in Ölt. He was in his midtwenties and had been there for ten years. But rather than digging deep mining shafts and painstakingly panning the gravel for gold, Buyanaa worked at night for only a few hours but was said to come home with his hands full of gold. Constantly surrounded by friends and many who would like to become his friends, Buyanaa was rumored to carry a secret that involved personal contacts with someone powerful, whether human or nonhuman. During my stay with Buyanaa, the key to his success soon revealed itself.

Early one morning Buyanaa came back to the ger with his regular workmate. They lit a fire in the stove and poured some wet soil into a small metal pan. They heated up the mix, then spread it out in their palms and blew on it softly to get rid of all the small

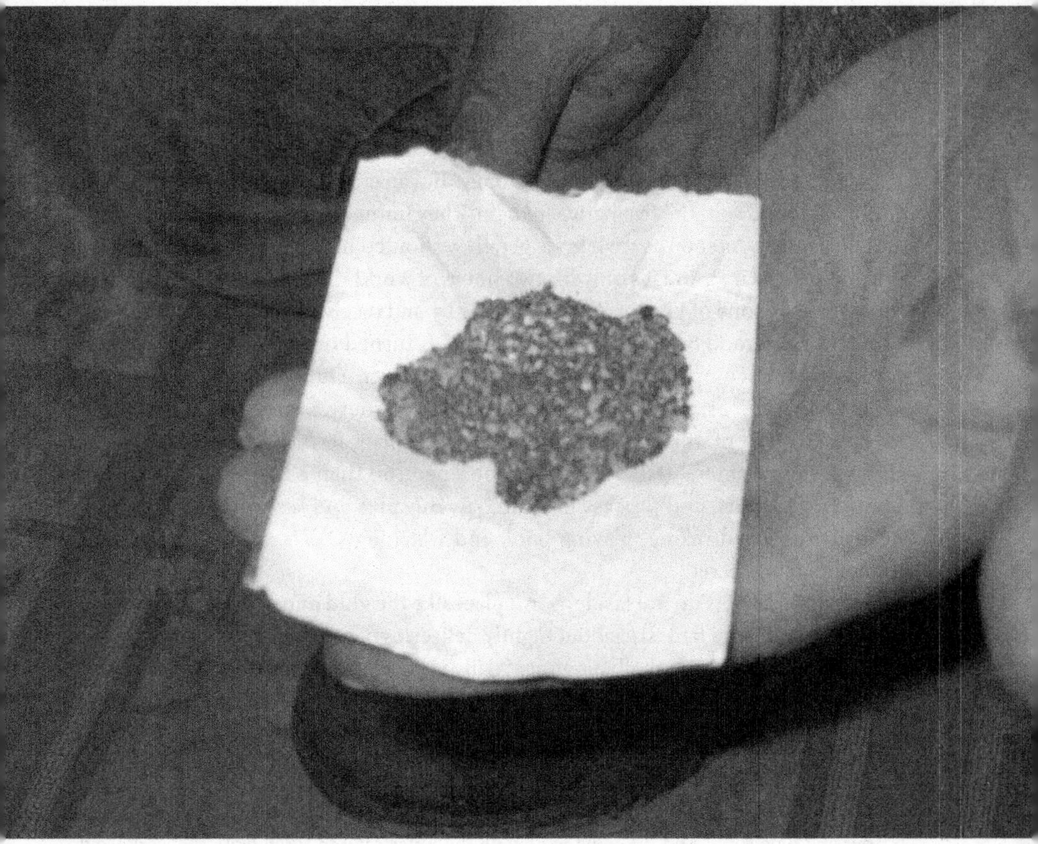

Figure 14. Showing the amount of gold found during one night's shift

pieces of stone. I was astonished to see how much gold there was. It was more than I had ever seen before in a ninja's palm (see figure 14). Having examined its purity, Buyanaa looked up with shining eyes, exclaiming, "Uyanga has the best gold in all of Mongolia. This is 99!"[7] After he wrapped up the gold in a torn piece of foil from a packet of cigarettes, we walked together to the nearby gold trader's ger. Despite the surging gold price, ninjas rarely hold on to their gold in attempts to negotiate a better deal. Given its perilous "black footsteps" (*har mörtei*), the unearthed gold is preferably passed on quickly to the traders lest misfortune follow.

The gold trader's ger was located in the middle of the muddy and potholed mining area. Inside, neatly arranged sales items such as cigarettes, vodka, and candy took up one side of the ger. The other side was lavishly furnished with a large flat-screen television and expensive "city-style" furniture. In contrast to the dirty surroundings outside, the ger was immaculately clean and tidy. Buyanaa handed over the gold to a young, smartly dressed woman with a wide-open money belt, packed tight with tögrögs. She quickly weighed the gold on a small electrical scale: 9.6 grams. Deftly counting the money, she passed 170,000 MNT to Buyanaa without any small talk or other interaction. Whereas Buyanaa was usually the center of attention, in the gold trader's ger he faced a palpable coolness and disregard. Clad in his torn, dirty clothes and standing in a pool of mud, Buyanaa suddenly did not look half the man he otherwise was among his ninja friends. Upon receiving the money, Buyanaa and his workmate put 10,000 MNT aside and split the rest between them. They immediately bought bottles of beer and fresh meat. As soon as we left the ger, they shouted in elation, "This is as much as one month's salary! And it's only for two hours of work!" This had to be celebrated. A few hours later one of Buyanaa's friends came by and received not only great amounts of alcohol and food but also the 10,000 MNT. It turned out that this friend worked for the mining company Erel as a dump truck driver. During his occasional night shifts he sometimes, if the managing director was off duty, passed one load of soil to Buyanaa instead of offloading it at the processing plant farther up the valley. Sifting through more mineral-rich and already loosened soil than a large mining team can unearth, transport, and process in a full day, Buyanaa and his workmate spent the next many days drinking, playing pool, and watching DVDs while guarding their money-making secret.

However, secrets do not last long in a place like the gold mines, and it was not long before I heard Ahaa fantasize about nightly deliveries of gold-rich soil. He had discovered Buyanaa's secret through his almost daily visits to the mines when selling milk to ninjas in the summer months. Given Ahaa and Ber's fear of hel am and their struggle to accept their position within Yagaanövgön's ail, Ahaa began to contemplate various alternatives, which he voiced in the following conversation with his mother:

> You know, in the mines people find gold every single day. There is plenty of it and it won't run out any time soon. And if I could work with the water cannon [*usan buu*], the work would be less hard, but the profits would be greater. We would make good money! I already know people there and I have heard about a guy who could deliver the gravel, so I wouldn't even need to do any of the digging! If I sell our animals, then I could buy my own water cannon with the money. And if I sold *all* our animals, then I could buy a water cannon *and* a car. If I had a car and learned how to drive, which shouldn't be difficult, right?—then I could set up a dirt taxi business [*shoroony taksi*].

His mother interrupted his train of thought:

> Stop talking like this. You're being stupid. You have many animals here, the herd will grow and your life is stable [*togtvor*]. But in the mining area, you will not always find gold. Your health can't take it either and your car will break down. And then what? Then what will you do? And what will you do with the money anyway? Many bad things will happen . . . you know that. What you have here is lasting!

Ninjas and herders view money earned from mining as something different from money earned by other means. Best kept in circulation, gold money should not be allowed to rest and definitely never stored, either in a bank account or at home.[8] The desire to make the cash circulate and not stand still may appear reminiscent of money-laundering practices. Arising from illegal activities, gold money, like the proceeds of a crime, requires transaction in order to appear of legal origin. But ninjas in Uyanga do not seem particularly concerned about the officially illegal nature of their earnings (see also Roitman 2006, 262). Instead they describe gold money as dangerous, heavy, and polluted—words that I have never heard applied to money earned through other means, including money earned from other illegal activities. Apart from highlighting specific qualities attributed to gold money, these terms are also central to a broader cultural logic of pollution (*buzar*).[9]

In the mines as well as on the steppe, pollution is an essential organizing principle that is both manifestly concrete and cosmologically abstract. In daily language the term *buzar* can be used in similar ways to the word *hog* (rubbish) and denotes something filthy or disgusting. It is commonly applied as an adjective yet is used only to describe particular nouns such as shoes, dogs, urine, feces, and human blood.[10] In distinguishing the dirty from the clean, it divides practically, discursively, and symbolically the upper from the lower, the male from the female, the orderly from the wild (Lindskog 2000). While these divisions are most strongly asserted within the domestic space of a ger, they are also apparent in many other aspects of social life such as posture, movement, and speech (Humphrey 1974; Lacaze 2006). If such divisions are not adhered to, the pollution may inflict harm on those present, and illnesses are thus often seen as caused by exposure to something polluted (see also Clark 1993). Although pollution can affect all members of a household group, women and children are considered particularly vulnerable (see also Humphrey and Onon 1996, 171). In order to minimize its harmful effects, people wear protective bracelets and necklaces, often in conjunction with birth year or Buddhist deity necklaces. In the ails of both Yagaanövgön and Nyambuu, women and children encircle themselves with burning juniper incense in the evenings in order to ensure protection from the heavy and dangerous pollution. Pollution is in this way not merely a descriptive label of impurity, nor is it limited to ideas about gold money. It constitutes a much wider organizing principle that underlies a cosmoeconomy that makes daily life possible.

Indeed, Uradyn Bulag (1998) has pointed out that far from the gold mines of Uyanga, buzar is also fundamental to postsocialist geopolitics and nationalist sentiments in Mongolia. He argues that there has been a historical change in how people engage with pollution. He notes that "foreign things were always regarded as polluting, but formerly the Mongols were able to absorb the pollution. Now it is not so easy, though they are trying" (ibid., 263). Emerging from a Soviet socialist past and with the spectral presence of China, a "chaotic situation" (ibid., 19) has arisen in which desires for and attempts at purification have magnified.[11] Rather than pacifying, domesticating, and incorporating that which is deemed polluted, these purification processes seek to reaffirm the fundamental divisions that are seen to offer protection.

In referring to money earned from gold mining as dangerous, heavy, and polluted, ninjas and their families therefore emphasize its distinctly harmful potential and its necessarily cautious handling. As ninjas dig deep shafts into the ground, they disregard fundamental cosmological distinctions, thereby allowing the flow of pollution from the lower into the upper domain, from the wild into the orderly. Moreover, the search for gold has led to the breakdown of fundamental boundaries that isolate the polluted substances of human feces and urine. Since ninjas constantly move their gers according to the establishment of new mining shafts, areas of open air defecation are no longer distinct and separate from areas of habitation and work. As Bilgee from my mining team noted when we walked together through an area of scattered toilet paper, "This place really is extreme [*ontsgoi*]!" With pollution moving across domains that are usually kept distinct, the mines have become intense places of impending misfortune. At times it affects individual miners, such as when a middle-aged ninja with a large abscess on his neck sought help from a village lama whose diagnosis confirmed buzar to be the cause. At other times, misfortune is transferred to the money miners receive as payment for the gold they unearth.[12] Capable of inflicting serious calamities, gold money, as a vector of pollution, can potentially affect anyone who becomes part of its subsequent circulation. Emerging from its conversion from unearthed minerals, money travels through the mines and across the steppe, enveloping its holders in an intense uncertainty about when and how a potential disaster might strike. Thus money earned from mining is far from being a form of wealth that is a straightforward and unproblematic antidote to the fears of envy and the spreading of hel am. As an inalienable currency that cannot be detached from its perilous origins in the mines, polluted money is accompanied by its own calamities.

The Purchasing Power of Polluted Money

In Nyambuu's ail, almost all his sons are involved in the gold rush. Since disagreements, if not violence, provide a common setting for their departures to the mines, their returns to the household on the steppe are often equally circumscribed by excitement and nervousness. As soon as they arrive, often accompanied by friends from their mining teams, Nyambuu's wife, Degidsüren, lights a butter candle on the altar, passes around some juniper incense, and recites a few Buddhist mantras for protection. She then serves some milk tea and begins cooking an elaborate meal. Eventually the ger is filled with conversation, banter, and laughter, at which point the visitors discreetly pass on gifts of money to Nyambuu, who just as discreetly puts it away. In a region where gifts, including money gifts, are generally received with much overt appreciation and exhilaration, ninjas' money gifts are received with minimal exchange of words and bodily gestures. As Degidsüren noted after receiving a share of her son's earnings from the mines, "Polluted money has black footsteps. Something will always follow."

In order to prevent calamities, families often seek to ritually purify the received money. By asking village lamas to cleanse (*ariulah*) the money or attempting to do it themselves, herders hope that the money will lose its destructive potential and become more like the currency that they see circulating elsewhere (see also Ferry 2002, 343). However, with the explosive growth in mining activities, the pollution in Uyanga's mining camps is now considered so rife that not even lamas are capable of eradicating the polluting potential of gold money. "It has become too heavy for us," one of the village lamas once lamented. Uncertain of the success of purification rituals, holders of cleansed money generally prefer to not spend it on durable objects. If people use it to buy a motorbike, they are said to be likely to soon have an accident. If someone buys a yak, a wolf may come at night and attack it. If people buy materials for their dwellings, it is feared that the ger may collapse at any moment. If polluted money, even when ritually cleansed, is invested in lasting objects like these, thus entering economic circuits of stagnation and permanence akin to that of pastoral wealth, the pollution can swell up and eventually harm its holder. In order to avoid the ramifications of "polluted wealth" (*buzartai bayarlag*), it is best to spend money earned from gold mining on fleeting subsistence goods, local entertainment, and, not least, alcohol. These transformative processes, rather than durable objects, are seen to leave no material traces of pollution (cf. Fiéloux 1980, 167, quoted in Werthmann 2003, 114n14; Luning 2009, 5–6).

When thousands of miners arrived in this otherwise quiet mountainous region, it did not of course take long before many shops suddenly opened in the village. Many of these shops cater specifically to ninjas, offering them plenty of fleeting subsistence goods and a wide selection of alcohol. Soylham opened her grocery shop in 2003, selling basic food items, alcohol, and general household goods. She sells more or less the same products as the other village shops, but she rarely complains about dwindling customer numbers, low sales figures, or the like. As she said, "We all sell similar products, so why would people not come here?" Whereas most of her customers are from the village, ninjas also frequently visit the shop. One day when I was paying her a visit, I noticed something unusual about her pricing.

A local schoolteacher had just bought a bottle of beer for 2,000 MNT. A few minutes later, a ninja entered the shop, asked for the same kind of beer, and was made to pay 2,200 MNT for the same bottle. The ninja did not object or try to negotiate the higher price. He willingly paid the requested amount and then left the shop. In contrast to many villagers who have small herds that provide some meat and dairy for daily sustenance, most ninjas are entirely dependent on market exchanges for their subsistence needs. Thus it might not be surprising that Soylham raised her price when dealing with a ninja customer. Indeed, another shopkeeper told me that since ninjas find their money easily, they can pay a little more. Or, as one of the local mechanics said, "Ninjas are greedy [*shunahai*]. They love money more than the environment." However, when I asked Soylham about the incident, she did not note the customer's dependency or easy access to money. Instead she began talking about the physical state of his banknotes. She explained that if the notes were new and crisp, she considered them safe (*ayuulgüi*). If, on the contrary, they were muddy and crumpled, they were probably from the mines and thus not as valuable (*ünetei*), and she would then ask for a little more (*ahiu*). According to her, it did not matter whether a customer was a ninja; what mattered was the physical condition of the offered money.

Another village shop was run by Pürevtogtoh, a heavyset elderly woman who was well known for her firm demeanor. A former herder, she had many close contacts in the countryside and received regular supplies of fresh dairy products and meat to sell in her shop. In the summer months she had a large iron pot full of milk, a bowl of dried milk curd (*aruul*), and a plate stacked high with clotted cream (*öröm*) on the shop counter. A fridge was packed tightly with various cuts of mutton and yak meat. One day she told me how the village had changed over the years and turned into a place where "everybody goes to the mines. There is now no one who doesn't go." I must have looked surprised because she continued, "Just look here. Look at my money! It is all worthless. It is all worn and dirty." She took a particularly muddy note from the wooden box where she kept her money. "Look, some of these notes [*tsaas*, lit. paper] are even taped together. The filth people live in. They mix everything. This money isn't valuable. It's bad." Rather than adjusting her prices according to the specific notes offered, Pürevtogtoh had chosen to instead simply operate with higher prices. If, however, someone offered her new, and thus for her more valuable, banknotes, she would then lower her price.

Despite the government-decreed official value of a note, shopkeepers in Uyanga posit a relation between the physical state of currency and its redeemable cash value. In this tense cosmoeconomy, the national "signs of authenticity" (Strassler 2009, 72), such as official signatures and watermarks, do not prevent decentralized revaluations. Nor do they seem to offer customers a feasible or persuasive ground from which they can challenge the shopkeepers' practices (see also Højer 2012, 40). According to Adam Smith, "the offering of a shilling, which to us appears to have so plain and simple a meaning, is in reality offering an argument to persuade one to do so and so" (Smith [1789] 1979, quoted in Meek, David, and Stein 1978, 352). In a region where money is approached as a "sensual substance" (Lemon 1998, 29), the argument that persuades people of a currency's value is not only anchored in its state authorship but also tied to its caked mud and crumpled paper. Although people try to brush off the dirt and straighten out the notes, they can rarely recover the lost value. In a village without local banks, holders of dirty banknotes appear to have little choice but to accept the shopkeepers' de facto redenomination of dirty money.

By altering the value of banknotes, shopkeepers in Uyanga effectively increase their profit margins on the goods they sell. This could be seen as a calculated financial strategy to boost earnings. In running their businesses, both Soylham and Pürevtogtoh do focus on their net profits, but they are also much concerned about their daily turnover (*güilgee*). One day Soylham proudly announced that she was now taking in around 300,000 MNT (250 USD) in sales every day. In order to increase this amount, she regularly goes to the village monastery, where she requests the recitations of appropriate texts from Buddhist lamas. Sometimes she also invites lamas to carry out rituals on her shop counter. Increasing the circulation of money is for her central to the running of her business. Pürevtogtoh also draws on the services offered by the local monastery to increase her turnover. But in her view, the very composition of customers in the village helps her achieve the desired circulation. As she explained, "Ninjas are poor people. They go to the mines and maybe only find a small amount of gold. So when they come here, they just buy one liter of milk or a small cut of meat. They never buy much. But that is good. What is good about ninjas is that they come often and only buy a little."

For Pürevtogtoh, each individual transaction contains a moment of circulation. Indeed, in Mongolian language there is a conceptual link between "selling" and "circulating." Rather than being predicated on a bilateral relation between a seller and a buyer, the commonly used verb *borluulah* can be translated into English as both "to sell" and "to put into circulation" (Bawden 1997, 59). By having many small rather than large transactions, Pürevtogtoh thus achieves an even greater degree of circulation. A desirable turnover is therefore not only about the amount of money acquired over the course of a day but also about the number of transactions.

This sense of circulation is captured in the English loan word, used in Mongolian slang, for the general term for trader: *chanj*.[13] A trader is someone who facilitates change, converting money into goods, goods into money. This emphasis on the circulation and reinvestment of money might be seen as a forceful expression of merchant capitalism (Marx [1867] 1967, 247–57), perhaps refracted through the free market ideology that has been espoused by Mongolia's various postsocialist governments. It can also be seen as a historical feature of classical gold rush frontiers and their initial transformation into mining towns with profitable auxiliary businesses (Rohrbough 1997, 197–216). The trade in Uyanga might very well give rise to similar processes of formalization, but the shopkeepers themselves do not associate their desire for high turnover and their practice of devaluing dirty banknotes with such capitalist ambitions.

Instead, they emphasize the dangers of holding on to monetary wealth in a region where polluted money flourishes. As Soylham said, "It is important for us to make a lot of money every single day, to have a high daily turnover. We need to transfer [*damjuulah*] the weight of the polluted money." In seeking to increase turnover, Soylham attempts to keep the polluted money in constant circulation. Rather than putting money aside for future investments, personal savings, or casual spending, she wants the heavy money that crosses her shop counter to move on and flow away from the shop (cf. Lomnitz 2003, 140). If it were to remain in her possession, resting in a bank account or stagnating in her purse, its potency could strike and cause misfortune. As soon as money comes into her shop, she therefore spends it on replenishing stock. With shelves already full of goods, her shop is often crammed with ever-more piles and boxes of new goods. Likewise, Pürevtogtoh's preference for many small, rather than large, transactions is also related to a concern about money's capacity (*möngööröö hirtei*) to be accompanied by trouble (*gai daguulah*). If one customer spent a lot of money in her shop, she feared it would be too heavy (*heterhii hünd*) and could not be passed on so easily. This undesirable stagnation may also occur when polluted money is accidentally brought into physical contact with unsold goods. At Mongolian markets, traders usually bless the taking and giving of goods by touching the remaining goods with the received money immediately after a transaction. But both Pürevtogtoh and Soylham warned against doing this with polluted money. They claimed that such a practice would risk contaminating the unsold stock with the pollution from the dirty money and make the whole shop a "magnet of misfortune" (see also Højer 2012,

38n10; Werthmann 2003, 113). Confronted with the heavy and dangerous pollution that accompanies the money, the shopkeepers have become nodal points in the transformation of polluted money into desirable, fleeting subsistence goods. Given their large and dependent base of customers, many of whom can pay a slightly higher price for the products, the shops in Uyanga have resorted to a careful balancing of redenominated banknotes with a high frequency of transactions.

The troubled origin of gold is thus present in both the moral evaluations and local considerations of the cash value of polluted money. Rather than being handled simply as the legal tender it is proclaimed to be and considered worth its stated exchange value, money earned from gold mining is effectively devalued in local exchanges. As shopkeepers scrutinize the physical state of banknotes and decide the extent to which they are crisp or crumpled, the materiality of currency indexes its past and makes visible potential threats it poses to its future circulation. As gold money becomes an objectification of morality, a previous holder's acts achieve solid and visible form. As such, it stands apart as a specifically unique and personal object that does not simply mobilize a generalized, all-pervasive enticing desire.

In a region where envy burgeons and hel am seems a constant threat, the incredible circulation of money that has been brought about by the gold rush might at first sight seem a solution, if not a blessing. Less visible and predictable than pastoral wealth, gold money evades easy quantification, thus rendering the basis for hel am attacks less potent. But as this chapter has shown, money earned from gold mining brings with it its own calamities. With its black footsteps, the heavy polluted money needs to be kept in circulation lest misfortune strike. Ninjas buy alcohol, and shopkeepers buy new stock. If allowed to settle, the pollution is said to swell, and bad things happen. Rather than evidencing the eradication of witchcraft and the emergence of a disenchanted modernity, the Mongolian mining boom has thus set in motion a profoundly uncertain future for many. And it is an uncertainty that now inhabits the very core of the cosmoeconomy: the symbol of value equivalence. Subject to revaluation, money in Uyanga is associated not only with the state but also with much more powerful and important local forces. As the prime interconnection between human and spirit worlds, the flow of fortune has been placed in jeopardy.

In their attempts to capitalize on the fortunes of gold, ninjas have come to rely on the support and assistance offered by Buddhist lamas at the village monastery. With most of their ritual services now directly related to mining, the lamas have become central to the local mitigation of impending misfortune and, not least, the circulation of gold money, as the next chapter will discuss.

5 Wealth and Devotion

ON A WARM August day in 2006, a large crowd gathered outside the village monastery to take part in the annual *Maidar Ergeh* (Sanskrit *Maitreya*) prayer ceremony.[1] Chanting Buddhist lamas were seated in the blazing summer sun, surrounded by dozens of makeshift market stalls, a motley swarm of lay worshippers, and giggling teenagers on horseback. The lamas swiftly spun their prayer beads, sounded their conch shell horns, and blew waves of juniper incense across the onlookers in anticipation of the future Buddha's coming. Discreetly retreating from his seat, Tögslam, a middle-aged lama who was a close friend of mine, joined me in the cool shade and initiated our conversation. "So many people have arrived today! I didn't expect this." I nodded in agreement as he continued:

> Our Uyanga has really grown over the years. In fact, too many people have moved here. . . . There are very few services available to them all. A new school has been built but there still aren't enough places for all the children to go to school. It is the same with the kindergarten. It needs to expand. And our hospital, it also lacks staff, not least properly skilled staff! At the moment there are only three people working at the hospital and there are so few beds. These are urgent issues! It is weird [*hachin*] because we were one of the first districts to get electricity. We also had *the* best well in the whole area! We became a super district [*süper sum*] already years ago. There was so much potential here.[2]

Tögslam paused and looked across the bustling crowd. An elderly, frail woman approached him, greeted him with head bowed and prayer beads in her joined hands. After a short exchange about her ill health, he passed on blessings and wished her protection from the black *lus* (see chapter 3). Quietly observing how she was helped back into her seated position in the grass by her relatives, he resumed his reflections on the current state of affairs in Uyanga:

> Since 1990 I have strongly supported our freedom [*erh chölöö*] to live and work. But people's living standards are going down. Now there is so much poverty, alcoholism, and poor health. As you know, our Uyanga is very rich in gold. It is also a place where people just seek their *own* gains and make a living by digging into the ground. But we can't blame nature for being rich in resources! Some people are doing offensive politics [*uls tör hiij dairah*].

For Tögslam, the "offensive politics" was not limited to corrupt politicians at various institutional levels in both the capital city and the countryside or to the scores of ninjas searching for Uyanga's gold, many of whom he had come to know personally over the years. It also implicated various nonhuman beings whom he considered as much part of "the living world" (*yertönts*) as its human population. Recognizing politics as involving more than just human agents, Tögslam feared that current offensive practices would have serious and urgent repercussions far beyond the human. Sometimes he expressed concern that Mongolia's gold rush was a sign of the world's moral degradation before it leaps into its "last age," as it has been described in Buddhist scriptures (see Nattier 1991)—an age in which Maitreya appears and guides us back onto the right path toward moral wisdom.[3] At a time "when rebirths were going downwards" (*daraa törööldöö dooshoo unaj baina*), Tögslam regarded himself and the other village lamas as important, even if inadequate, mediators who could offer attention and care to all parties so that life could be as wonderful (*saihan*) and peaceful (*taivan*) as possible for all beings (figure 15).

Figure 15. Village lamas at the Maidar Ergeh celebrations in Uyanga

As the village lamas attempt to mediate the offensive politics, they have become centrally involved in the cosmoeconomy of the gold rush. This chapter examines how they make sense of this involvement and position themselves vis-à-vis the circulation of gold money. Rather than offering an account of Buddhist theology as a distinct kind of abstract epistemological orientation separate from economic life (see also Scott 2009), I consider here a Buddhist ethics of self-transformation in conjunction with concerns about Mongolia's current mining boom. In previous chapters I have described how the extraction of gold is seen to transgress fundamental taboos (*tseer*) related to the land and upset powerful spirit beings, especially black lus. I have also shown how people live amid the invisible dust of *altny chadvar* that can pull them against their own will and the *altny gai* that can be used in malevolent ritual attacks. I have further shown how the mines are considered intense (*ontsgoi*) places where pollution (*buzar*) flourishes, affecting those washing dirt (*shoroo ugah*) and the money they earn. As Uyanga's lamas attempt to address these multiple flows of misfortune through ritual action, this chapter focuses on the dynamics of the moral degeneration that they see burgeoning at present—a degeneration that they associate with many people's mistaken understanding of freedom.

The Politics of Compassion

Mongolian Buddhism is generally associated with Mahayana (Sanskrit *Mahāyāna*) Buddhism and its Tibetan variations, while also incorporating various local religious traditions (see Heissig 1980; Wallace 2015). Within this confluence, the notion of freedom is intricate and multilayered. In the classical Buddhist teachings, "freedom" refers to the ultimate escape from the deluded nature of samsara (Sanskrit *saṃsāra*) and the attainment of enlightenment. The Sanskrit word *saṃsāra* literally means "continuous flow" or "to perpetually wander." It refers to the flow of existence as one passes through birth, life, death, and rebirth. The Mongolian word is *orchlon*, and it approximates to a quasi-place name that denotes "this world" as opposed to the much worse "underworlds" or "hells" (*tamyn oron*) or the distant world of bliss inhabited by gods. It is a realm of existence that is at once abstract and immediate, speculative and concrete. It is closely associated with the view that beings continue to be born and reborn into any of these realms depending on the karma (*üiliin ür*, lit. result of actions) they have accrued during life. As such, orchlon underscores the stakes involved in ethical action in this current world. It is through contemplating our actions, caring, and taking care that we can leave it behind. And as a result of this, the emphasis on ethical self-formation does not abandon the realm of politics. The ethical and the political are not separated or compartmentalized but rather exist in unbroken continuity (see also Mahmood 2005). The project of freedom is thus to bring to fruition the potential that most humans have within them to transcend their current bodily form and no longer be caught within this worldly cycle of rebirths.

When one of the lamas in Uyanga described this sense of freedom to me, he referred to a famous twelfth-century Tibetan text entitled *The Jewel Ornament of*

Freedom (Tibetan *thar pa rin po ch'e rgyan*). It provides an account of how to follow the Mahayana path to freedom: "On account of this human existence endowed with freedoms and assets, there is the ability to give up non-virtue, the ability to cross samsara's ocean, the ability to tread the path of enlightenment and the ability to attain perfect buddhahood" (Gampopa 1995, 21).

For humans, freedom is described as a precious jewel of liberation—a potential so pure, complete, and timeless that it should be treasured and rejoiced in accordingly. According to the Dalai Lama (1994, 4), the desire for this freedom constitutes the shared foundation for ethical sensibilities among all beings. In his view, "All beings are equal in the sense that all have a natural tendency to wish for happiness and freedom from suffering." Since all beings are subject to the laws of karma, escaping samsara and attaining enlightenment are goals not only for humans but for all beings. However, not all beings are considered equally positioned to achieve the ultimate goal of freedom from suffering. Beings born into lower realms, such as hell beings, hungry ghosts, and animals, are said to be driven by their delusive passions and desires. This makes them unable to rein in their untamed minds. Humans, on the other hand, are endowed with a precious consciousness that makes it possible for us to perform actions of mindfulness, restraint, and compassion. In contrast to other beings, humans have thus obtained what the Dalai Lama describes as "the best form of existence for the practice of the Dharma" (ibid., 39). As a result, they are under a particularly strong moral imperative to realize their unique potential for freedom.[4]

For the lamas in Uyanga, however, gaining enlightenment and reaching nirvana (Sanskrit *nirvāṇa*) does not lie within their purview. They consider themselves so-called low sutra lamas (*baga hölgönii lam*), entangled in the present and primarily concerned about helping other people with their pain and suffering. After a busy morning at the monastery, one of the lamas introduced himself in the following way:

> I am the kind of lama who people come to for help when someone is sick. For example, a child is crying and can't sleep, a woman is having difficulty giving birth, or a man is experiencing problems with his eyes. So I recite relevant mantras [*tarni*] and blow on the painful area. I might hand them some blessed water or ask them to light some incense at home. I give blessings [*yerööl tavih*]. I am that kind of lama.

Whereas other Buddhist traditions emphasize the liberation of the self as the goal of religious devotion, the Mahayana school emphasizes the liberation of all beings as the central motivating factor. The Buddhist scholar Jonathan Landaw attributes this collective emphasis to the fundamental view that "we are not the only ones who experience suffering and dissatisfaction; all other living beings share in the same predicament" (quoted in Yeshe and Zopa Rinpoche 2012, 16). In cultivating compassion—that is, the altruistic desire for other beings to be free of suffering—Mahayana Buddhism foregrounds exemplars of compassion: the so-called bodhisattvas (*bodsadvaa*, Sanskrit *bodhisattva*). These beings have reached a state of enlightenment but have decided to postpone entering nirvana in order to use their merit to release from suffering all who pray to them. Although the lamas in Uyanga seek to help other people with their pain and suffering, they are far from reaching the blessed state of the bodhisattvas. Instead they consider themselves earthly beings who are still held in this immediate world of orchlon.

As is common for small and little-known monasteries across the Mongolian cultural region, none of the village lamas in Uyanga have taken the vows of monkhood (*gelen sahiltai*) (Bareja-Starzynzka and Havnevik 2006, 221). According to one of the lamas, "a very small part of him" was thus committed to accept the commandments of the Buddha. Most lamas are married and live with their wives and children. They drink alcohol, smoke cigarettes, and eat meat. In contrast to the so-called high sutra lamas (*deed hölgönii lam*) in Ulaanbaatar, the lamas in Uyanga are not obliged to accept a strict monastic discipline and indeed rarely practice deep meditation, philosophical discussion, or prolonged isolation. They rarely go on pilgrimages or study under renowned Buddhist masters in Inner Mongolia or India. They are village lamas who are emphatically caught in the present and its many difficulties. Freeing themselves from this world is thus not within their reach, nor is it their call. As a younger lama put it, "I am a lama who can't set the goal of becoming pure [*ööriigöö ariusgah*] but I can instead pray for the good of every living thing."

Since the reopening of Uyanga monastery in 1990, only a limited number of lamas have continued to pursue their monastic practice. Many who have undertaken monastic training over the years have since left the monastery and joined the lay population. An elderly lama explained,

> After reaching the age of twenty, some lamas tend to get married and have children. They face a situation where money is required and they realize that the salary they get from here is not enough to take care of a family. They also learn that life as a lama is not that easy, that the life of meditation might not be that comfortable. They then go to the countryside to become herders or begin to work in the mines. They start to think of new ways of making money and so they leave. It might be said that they have left because of greed [*shunal*].

Although many have left the monastery, one of the most experienced lamas in Uyanga has been there since its reopening.[5] Budlam, who hosted me for months, comes from a family of lamas: his father specialized in Buddhist philosophy (*choir*) and his father-in-law in Buddhist medicine (*mambe*). After years of working as a math teacher at the local school and later as an accountant for a small company during the socialist period, Budlam joined the monastery in Uyanga at the cusp of independence. During any given day, he was busy attending to the needs of local people. Long queues formed in front of his desk at the monastery where he did readings (*unshlaga*) for several hours every day. Once he returned home, there was usually someone requesting him to carry out a personal reading. Sometimes he performed readings in his home; other times he would go to people's gers. Since he was considered the most capable and knowledgeable lama in Uyanga, people often asked specifically for him at the monastery—even if it meant waiting around for hours, if not days.

The Mining Boom

Part of the reason why Budlam was so busy was related to the mining boom. As one of the lamas described it, "Mining is like having a thorn pressed into your hand. That's how much nature and the land hurt." Another lama elaborated, "When people mine for the resources of the land, they are breaking the energy of the landscape [*baigaliin energii evdeh*]; they are breaking the natural law concerning nature and the universe. They are destroying the living things and doing something that causes wrath [*süitgej baina*]."

As described in earlier chapters, mining in Uyanga relies on the central use of water, and it is therefore, above all, water lords known as lus that are considered angered. These beings are in many ways reminiscent of *nagas* (Sanskrit *nāga,* Tibetan *klu*) in Buddhist cosmology. The Sanskrit word *nāga* means "snake" or "serpent," and nāgas dwell in rivers, lakes, and other water sources. They are depicted as having their own society or kingdom under the earth and are passionate beings that will strike back at offenders (Nebesky-Wojkowitz 1956; Tucci 1949, 723). Nāgas are further depicted as creatures with the torso and head of humans and the body and tail of a snake. Moreover, they are said to be able to assume human form at will (Donald Lopez, personal communication). The religious scholar O. H. Sühbaatar (2001) collected such examples from various parts of Mongolia. However, in Uyanga I did not come across descriptions of anthropomorphic features or abilities in lus. Although lus and humans are regarded as inhabiting interconnected realms, their bodily manifestations are here generally thought to be distinct.

When traveling along redirected rivers, stagnant tailing ponds, or dried-up riverbeds, herders often lamented the disregard (*toohgüi*) that they felt ninjas had shown local residents, whether human or nonhuman. And when the Ongi River eventually turned deep red from the heavy mineral and sedimentation pollution from mining (see also chapter 2), it acquired the colloquial name of Red River (Ulaan Gol). This name referred not only to the changed color of the river but also to the "red sentiments" that herders and lamas commonly ascribe to the thousands of ninjas working in the area. In the Mongolian language there are many expressions that associate the color red with negative feelings of anger and violence. Those that I heard used most often to describe ninjas included *ulaan galzuu* (red rage), *uurlaj ulaih* (to become red with anger), *ulaan üzsen araatan shig* (to be like a wild beast that has seen red), and *ulaan herüülch* (a red, quarrelsome/argumentative person). As an *ulaan herüülchid*—literally a "red, quarrelsome people"—ninjas are seen to instigate angry confrontation and conflict by refusing to care about other beings.

Many of Uyanga's lamas also point to another reason that lus have become particularly affected in and by Mongolia's mining boom. As Budlam recounted, "Our people never really liked to take a lot of gold. It is a very special and precious treasure that is one of the 'nine jewels' [*yösön erdene*] connected to the gods."

The nine jewels include pearl, coral, turquoise, lapis lazuli, mother of pearl, steel, copper, silver, and gold. Historically these gems were sometimes ground up and used as pigments to write Buddhist texts, representing the gem-like teachings of the Buddha (see Zhengyin 2003, 618). Of these nine jewels, gold is considered an unmatched treasure that is held (*barih*), if not withheld (*tatgalzah*), by lus. Unlike many other metals, gold is fiercely guarded by these beings. As in the material collected by Réne de Nebesky-Wojkowitz (1956, 253) on Tibetan protective deities known as "treasure guards," lus are said to have adorned their abodes with gold. Staircases, walls, and floors are all made of gold, as is their armor if they have any. They might be protecting themselves with gold shields or brandishing swords made of gold. They might drink from gold cups or curl around enormous gold nuggets. In order for humans to obtain gold without stealing from lus, it is therefore necessary to first try to persuade lus to part with their treasure.

Desiring a metal that is so closely guarded and so destructive to extract, ninjas often come to the lamas for help. Sometimes they ask them to carry out rituals that will calm the angered spirits.[6] Other times they ask the lamas to relieve the physical pain they experience from working in the mines. If ninjas have not been able to find much gold, they also ask the religious specialists to cajole lus into sharing fortune (*hishig*) with them. Or, if they are about to open a new mine, they invite several lamas out to the "land of dust" to conduct a large prayer ceremony. It is not only ninjas who draw on the lamas' expertise to carry out rituals. Villagers, herders, and others also make frequent visits to the monastery and request ritual assistance. As noted by Johan Elverskog (2006), Mongolian Buddhist practices often center on ritual efficacy, and there is much liturgical emphasis on the reading of texts, making astrological predictions, and conducting personalized rituals. And for the lamas in Uyanga, ninja miners and company bosses have now become their most frequent and loyal visitors by far.

Yet, according to most lamas, the ninjas are not *really* concerned about the impact of mining on spirit beings. They are not *really* troubled by the pain they inflict on other beings in their disregard of taboos. In response to my questions about the high number of ninjas visiting the monastery, they insisted that this was not a case of an economic boom accompanied by intense religious revitalization. Nor was it for them evidence of an "occult economy" in which human desires are mediated by concerns about angered spirits (Taussig 1980; Comaroff and Comaroff 1999; Geshiere 1997). Nor, for that matter, did they view it as a "spiritual economy" in which the geist of religion works in tandem with the goals of economic maximization (Weber 1904; Rudnyckyj 2010). Much to my initial surprise, in their view the number of ninja visitors was rather evidence of the longevity and popularity of Marxist materialism. They regarded the current pursuit of gold as an example of how their fellow citizens think that the land and its natural resources are simply there to be exploited. In the words of a younger lama, "They think that what truly exist are only the things that are materially real, that everything else is empty." Seeing the universe in these terms, ninjas are presumed to come to the lamas not because they recognize the existence and suffering of spirits but because they are greedy for money. They just want the lamas to make it more likely for them to find gold by doing whatever rituals might be needed. If they really were concerned about inflicting suffering on others, they surely would have given up mining a long time ago. As Budlam said,

> Today, ninjas come with swollen legs and arms. When I blow on their body it heals again. Then they think everything is sorted out. But that's not true. The truth is that they are destroying the living world [*yertönts*]. They just don't know that this is a living thing. They were taught to believe that the environment is lifeless. But it's not. This whole universe is a living universe and can't be treated as separate from its humans and animals, from its worms and insects, from its water and air. All these together are one whole thing.

Perhaps not surprisingly, few ninjas would agree with Budlam's rendition. Instead they remark that their frequent visits to the village lamas reflect precisely their recognition of and concern for spirits. The fact that they do not change their ways or give up mining is, to them, a result of the limited alternatives available. In Uyanga, their most likely alternative is pastoralism. But as chapter 1 has shown, the herding economy is centrally structured around patriarchal power relations that can prove tense and undesirable for many household members. Although it has been the traditional backbone of Mongolia's economy for centuries, pastoralism is not necessarily considered a desirable or viable option for many ninjas.

In an article on the moral apocalypse experienced during postsocialism, Caroline Humphrey (1992) describes how Mongolians have become "dislocated selves" faced with the challenge of creating anew a "truly Mongolian" moral society. As she writes,

> Soviet ideology was taken up almost more sincerely, more naively, more brutally than in the USSR itself. In the 1930s the Mongolian government destroyed every single one of the 700 Buddhist monasteries in the country and killed tens of thousands of people, annihilating all that was best and most sophisticated about native Mongolian culture, philosophy, and art. . . . A feudal society permeated with religion at all levels was abruptly replaced by a European, atheist ideology, predicated not on a model of the past but on the modernist development of the present. The moral authority of the socialist period was based on a vision of a future society, which was to be egalitarian, industrialised, and single-minded. (ibid., 375)

The socialist attempt to eradicate anything traditional, anything distinctly Mongolian, and certainly anything religious, has in many of the lamas' eyes been highly successful—probably more successful than ever intended (see also Kaplonski 2003, 2008).[7] And indeed, the radical political initiatives that were undertaken in the name of socialist egalitarianism have, in their view, ended up paradoxically producing individuals who have lost their concern for precisely the collectivity. In a vein characteristic of the elder generation of lamas, one of the oldest lamas in Uyanga views the historical transition this way:

> We, the Mongolian people, used to be a nation that lived for others. But after seventy years of socialism, Mongolian people only live for themselves. If a stray dog starves, no one will now give him food. If a drunken man sleeps on the ground, no one will now give him shelter. No one will care. Mongolian people nowadays don't take pity on others. When did that good thing of Buddhism, that thing that teaches people to live for others, get lost? During these seventy years of socialism, when everyone started to live for themselves only, for making his own life better, for getting his own salary, for desiring to have a comfortable life and striving for nothing else, people of our generation stopped dedicating their minds for others. They completely stopped caring for others. Mongolia is being destroyed by itself, by its own people.

This elderly lama saw a cynical materialism pervading Mongolian society, inflected and promulgated through contemporary capitalist ideology (see also Humphrey 2002, 84).[8] The self, which was freed from "spirituality" under socialism, is today on the hunt for money, resorting to whatever means might be available. If this means digging deep holes in the ground and redirecting rivers, people will do it. If it means selling the potent metal of gold that so easily blinds its holders, luring them into bottomless greed, they will do it. The freedom that Mongolians aspire to, according to many lamas in Uyanga, is emblematically self-centered. It is free of political ties, free of moral obligations, free of environmental restrictions. In this view, socialist atheism has been elaborated and extended into a fundamentally free individualism that is conducive to neoliberal ideology as promoted in Mongolia's current mining boom.

Buddhist Environmentalism?

The lamas in Uyanga are not alone in making these critical remarks about the present state of affairs. Influential Buddhist practitioners across the world are responding to various contemporary problems, whether it is the Dalai Lama working toward the political recognition of Tibet, Thailand's Sulak Sivaraksa criticizing Western approaches to economic development, or Thich Nhat Hanh calling for a return to a more mindful engagement with the present, to name but a few. Whereas some see the environment as an instructive and exemplary teacher, others embrace a protective and defensive stewardship of nature and its resources. In recent years Mongolian lamas have collectively sought to address the needs and demands that have resulted from mining. Apart from undertaking various kinds of environmental awareness projects, many monasteries also collaborate with international Buddhist movements, development organizations, and the government. They print books and brochures on environmental issues, undertake reforestation projects, and organize environmental training courses for the lay population. According to the Centre of Mongolian Buddhism, based at the country's largest monastery, Gandantegchinlen, located in Ulaanbaatar, this involvement of Mongolian Buddhists in environmental advocacy work adheres closely to the teachings of the Buddha. Its statement reads as follows:

> We need to live as the Buddha taught us to live, in peace and harmony with nature, but this must start with ourselves. If we are going to save this planet, we need to seek a new ecological order, to look at the life we lead and then work together for the benefit of all; unless we work together no solution can be found. By moving away from self-centredness, sharing wealth more, being more responsible for ourselves, and agreeing to live more simply, we can help decrease much of the suffering in the world. (Buddhist Statement on Ecology, quoted in Chimedsengee 2009, 4)

The Dashchoilin monastery, also located in Ulaanbaatar, is commonly regarded as particularly efficacious in mitigating the damage caused by mining. Before a new mine is opened, lamas at Dashchoilin are often consulted. With reference to instructions contained in Tibetan and Mongolian texts, a mining site is determined to have relatively good or bad properties, just as a given day is regarded as more or less auspicious for "the breaking of the soil." Speaking at the Northern Buddhist Conference on Ecology and Development in Ulaanbaatar during my fieldwork in 2005, one of its lamas described how more and more mining companies were consulting them to inquire about auspicious start dates for their projects and to request "appeasement ceremonies" (*argadan örgöl*) addressing in particular angered lus. He believed that company directors were increasingly turning to them for assistance "because

Figure 16. Lamas conducting a ritual at a mine. Credit: Tim Franco.

company profits are decreasing and because families in the area are perceived to be experiencing harm from pollution and other similar causes" (Chimedsengee 2009, 22). Knowing both the companies and the affected families well, he emphasized that he offered ritual services and personal advice to miners and company bosses out of his compassionate desire to help mitigate human and nonhuman suffering (see figure 16). However, at a monastery where apparently 80 percent of Ulaanbaatar-based mining companies come for ritual services, it would seem that a new and substantial mining-related clientele has emerged. Given this high proportion of mining companies actively seeking the assistance of lamas in the mining boom, the religious specialists seem to facilitate, if not indeed legitimize, practices that they themselves consider offensive. As they carry out appeasement ceremonies and recite texts that encourage lus to pass their treasure to humans, are the lamas not also involved in the offensive politics for which they criticize the ulaan herüülchid of ninjas? Amid the project of self-transformation, are they not evidencing a way of being that is also fundamentally free and self-centered?

According to the head of Gandantegchinlen monastery (*Hamba lam*), such a view separates actions from the soteriological motivations of the involved lamas. He argues that it is because the fate of humanity is inextricably interwoven with and dependent on the fate of the environment that the Centre for Mongolian Buddhism encourages ritual involvement. Drawing on the belief in rebirth, he explains that since humans can be reborn as any animal, including insects—or even as trees or water—we must respect these elements of life. Prior to our present existence, we have known the suffering experienced by these beings because we have, at some point, lived a life in that body. Harming any being is dangerous for humans who may be reborn as any of these in the future. In this view, mining is a form of violence, inflicting on others the pain and suffering that we are all intimately familiar with. It entails a profound ignorance of other beings, especially those who do not have a human body. If lamas were to no longer conduct appeasement ceremonies and other rituals, they would then, in his view, be presuming the environment to be empty (*hooson*) and humans to be free (*chölöö*). He has therefore called on all Mongolian monasteries to "support and work closely with environmental organizations, encouraging above all the participation of the religious community" (Dudley, Higgins-Zogib, and Mansourian 2005, 8).

When I talked to Budlam in Uyanga, he drew on a slightly less conventional Buddhist critique of extractive industries. Attacking the ideological framework that justifies practices such as mining, he sought to prove how Marxist materialism and its celebration of the atheist individual misconstrue the relationship between humans and the environment. As he said,

> This atmosphere, this vast seemingly empty space called the universe, is not empty. It truly exists in the same way as material objects exist. All beings, such as animals and humans, trees and rivers, emit energy [*enerji*] through their own minds and actions.[9] They emit electricity, heat, and energy. As long as they think of good things and do good deeds, they disseminate energy that causes other living beings to also think of good things and do good deeds. But if they think of bad things and do bad deeds, then that energy also gets spread around them through the environment and into the atmosphere. This bad energy causes others to suffer and degrade. This energy always emanates from beings, especially from humans as they have a higher level of rebirth.
>
> Now, today, when we use mobile phones to get connected to each other, who is connecting us? How can we talk to someone in America, directly? The great energy that is constantly being emitted from living beings in the vast space of air, that positive and negative energy, that plus and minus energy, that energy of hot and cold, that's the only thing connecting us. This universe is one whole sphere of energy. Marxists say nature is empty, but it's not. That's why we can talk on mobile phones.

Budlam here draws on technology as material evidence for the existence of beings that we otherwise might not recognize and sometimes not even see. In this sense sentient life is everywhere around us. Its usual invisibility, however, does not justify our ignorant actions. If we mistakenly think that we are free and independent, we just need to cast a quick glance at the ways in which we now communicate across great distances. Incorporating technology into his view of the interdependence between humans and nonhumans, Budlam questions the human-centeredness of Marxist materialism and questions the supremacy it has so successfully lent to the material.

In Buryatia, Buddhist monks similarly draw on science and technology in attempts to assemble persuasive evidence of Buddhist understandings of being. When the Bolsheviks established the Soviet state in 1927, a high-ranking Buddhist monk known as Etigelov settled into prolonged meditation (Bernstein 2011; Quijada 2012). Having chanted tantric texts for the dead, he left the corporeal plane of existence and was subsequently buried. In accordance with detailed instructions that he had prepared before parting, he was exhumed in 2002. He was reported to have emerged still seated in the lotus position, and scientists stated that his miraculously preserved body appeared as if dead for only days rather than decades. The scientific testing underscored the common view that "Etigelov's lack of decay is not merely a miracle; it is a scientifically proven miracle" (Quijada 2012, 145). Linking his bodily state to Buddhist notions of transcendence, the monastic community drew on science to prove their faith, thereby reconciling the long-standing dichotomy between science and religion that was so dominant in the Soviet socialist period (see also Luehrmann 2011). After years of distress brought about by a failing state, people's desire for treatment and healing altered the relationship between these different kinds of evidential regimes, allowing science and religion to now become mutually instructive and supportive.

As one of the first in Uyanga to be equipped with a mobile phone, Budlam refers to technology as evidence for his understanding of the world and the ways in which all living things are mutually interrelated (see also McMahan 2008).[10] For him, this is not only a philosophical reflection on the state of the universe but also an understanding with powerful ethical and political implications. If we are all part of a vast, interdependent network of being, our actions can have profound implications for others. Conversely, lacking any inherent self-existence, actions carried out by others can also have profound implications for us. In a world where we are all constituted by our interactions with other beings, humans are far from the only agents on the political scene. And whether or not Budlam and other lamas seek an active role in the gold rush, they know they will be affected by the inflicted pain. As Lama Zopa Rinpoche suggested on a recent visit to Mongolia, "We have created the situation in which we find ourselves, so it is also up to us to create the circumstances for our own release." For the country's lamas, it is therefore pertinent to focus on easing the suffering inflicted on others.

The Matter of Gold Money

With ninjas frequently visiting the monastery, the lamas are aware that great quantities of gold money flow through their institution. Resulting from "digging for the resources of the land," the money is subject to much debate and disagreement among the village lamas. But, as Rachelle Scott (2009) has demonstrated in her ethnographic and textual analysis of the Dhāmmakaya Temple in Thailand, this is not because wealth is inherently troublesome within Buddhist practice. Since many lamas in Uyanga have had to forgo a monastic life partly because of the low income offered by the monastery, issues of wealth are recognized locally as vital to their involvement in Buddhist monastic practice. Often confronted with insufficient remuneration from the monastery, is this postsocialist generation of lamas drawing on environmental discourses while seeking to profit from an afflicted cosmoeconomy?

Perhaps not surprisingly, many of Uyanga's lamas object to such a reading. They explain that the monetary aspect of rituals is not that straightforward. As a younger lama grumbled after a particularly busy service, "All the big mining bosses came by here today. But it doesn't matter how many readings we do during our meeting [*hural*]. We still don't get any more money."

At Uyanga's monastery there is a set price for rituals, paid directly to a "lay accountant" (*nyarav*). The accountant stores the received money in an old wooden box, which he keeps together with a ledger of financial transactions. One day he showed me that the box was piled up high, right to the lid, with the soiled money notes that are so common in Uyanga. As he glanced at the money, he commented that "this almost isn't money anymore!" From these earnings, the accountant hands out a salary every month to each lama depending on his seniority. Approaching seniority largely in terms of age, the youngest lama receives 60,000 MNT per month (approx. 50 USD) with a sliding scale up to the most senior, who receives 100,000 MNT per month (approx. 83 USD). Since only few lamas observe even the basic code of ethics enshrined in the Five Precepts (Sanskrit *pañca-śīlāni*), the older the lama, the more senior and more highly paid he is likely to be—almost regardless of his abilities to study scriptures, engage in devotional practice, or cultivate higher wisdom. This approach to seniority, grounded largely in age, is in line with the broader regional practice of patriarchal social organization as described in chapter 1. The salaries are more or less fixed, so in times of greater ceremonial activity, the extra income is not necessarily distributed to the lamas themselves. It might instead be saved for quieter months or subsumed into the substantial maintenance budget. "It is expensive to keep the monastery operating," the accountant explained to me. In the old days the cost of basic food and ritual supplies was covered by the laity through alms, but today such networks of support have largely disappeared. Although the lamas might not receive a higher salary from their ritual activities during *hural*, the accountant explained how the monastic institution was certainly benefiting from the mining boom.

Beyond the set price for rituals, visitors to the monastery also often present offerings (*tahil*) to gods and, depending on the circumstances, to the lamas individually. Elaborate arrangements consisting of sweets, dried milk curd, and silk scarves topped with crisp new money notes are presented to the lamas, recognizing the position of the higher-ranking lamas with more extensive offerings. It is not rare for such offerings to include the highest denomination notes of 20,000 MNT (approx. 17 USD). During my stays with laity living in the mines and on the steppe, I noticed that people kept "new money" (*shine möngö*) separate and used it particularly for gift-giving purposes, including it in gifts to friends, relatives, lamas, and others. Sometimes people acquired this new money from banks when visiting the regional capital or Ulaanbaatar. Or they received it as part of gifts given to them by others. These monetary offerings constitute a substantial additional income for lamas beyond their monthly salaries. This is especially the case in Uyanga, where unfortunately there is reputed to be at least one fatality a day from the gold mining. During the intense transition period of forty-nine days known as *bardo* when the soul (*süns*) of the deceased travels between lives, relatives often present additional offerings and request additional readings from lamas. By thus "making merit" (*buyan hiih*), relatives can aid the journey of the parting soul. These sad occasions take place regularly and involve extraordinary levels of generosity. In a region of relative poverty, some of Uyanga's lamas have now become noticeably well-off.[11] In addition to the big bosses of the illegal gold trade, discussed in the following chapter, these lamas are now among the very few who own Land Cruisers and computers, invest in local businesses, and go on international holidays.

Yet they dismiss the moral significance of this wealth. The lama administering the Uyanga monastery (*da lam*) pronounced, "Material wealth can bring much misfortune [*gai*] and malicious gossip [*hel am*] among the laity, but not among us. This is because it only matters to the greedy mind [*shunal setgel*]."

As one of the wealthiest lamas in Uyanga, the da-lama insisted that to criticize him for his material wealth was a reflection of people's own greedy minds rather than a shortcoming of his. Since he did not suffer from such ignorance, he could own multiple cars and houses without having these possessions "pull at him from behind." And when I broached the topic of sharing his wealth with the much poorer residents of Uyanga, he reasoned that sharing his prayers was much better—"I think what is best for the Mongolian people is the philosophy of Buddhism, not wealth and possessions."

I wondered whether, in line with the local monastic practice of addressing and relieving suffering, the lamas' reluctance to share their wealth was perhaps also grounded in a compassionate concern that the muddy money notes that form the basis of the monastic savings and the lamas' monthly salaries could cause harm among the laity if shared—just like the polluted money described in the previous chapter. However, I soon learned that the village lamas regarded money in ways that differed significantly from the views of the lay population. They often outright rejected the view that there was anything *materially* problematic with or even significant about gold money. Although gold money was considered distinct by its physical dirt and crinkles, it was not necessarily deemed polluted (*buzartai*) or indeed any less valuable (*ünetei*) in the ways that the ninjas, herders, and shopkeepers in the previous chapter thought. Yadamsüren, a middle-aged lama, voiced an understanding that I often heard among Uyanga's lamas: "Crinkled and dirty money notes don't have to be understood as having the deep meaning of pollution. Maybe the money notes have just become old because people have used them again and again in their lives. Or maybe the money notes just got dirty accidentally."

However, not all the village lamas agreed with Yadamsüren's rejection of the significance of money's materiality. Tömörbaatar, a lama in his early fifties, maintained that the paper of money is far from trivial and inconsequential. In his view, the material matter of money is invested with deep meanings and as a result ought to be handled respectfully by people. As he explained,

> Money notes should be treated with much care and respect. They carry the national flag and pictures of our Chinggis Khaan and Sühbaatar, all representing our independence.[12] But the

> Mongolian "soyombo" symbol is also printed on them. According to the teachings of *burhan* religion [Mongolian Buddhism], it is a very special and rare wonder. It has a really beautiful meaning. That's why money notes should be treated respectfully [*hündetgeh*].

The symbol that Tömörbaatar refered to is part of the script known as *soyombo bichig* (see figure 17). The word *soyombo* derives from the Sanskrit *svayambhu*, which means "self-existing" or "self-evolving" (Atwood 2004, 518). The script is accredited to the Mongolian scholar monk Öndör Gegeen Zanabazar (1635–1723). According to Tömörbaatar, Zanabazar did not methodically or systematically invent the script. Instead, he was said to have noticed some vague traces of the symbolic signs suddenly appearing across the sky one night in 1686. Having this vision, Zanabazar instinctively knew its significance and memorized its configuration. This was the beginning of a unique script that today survives in inscriptions on monastic buildings, stupas, and prayer wheels throughout the country.

In the script, the soyombo symbol is the character that is used to mark the beginning of short Buddhist texts (see figure 18). It is highly intricate and composite. For Tömörbaatar, its elements symbolically represent the path to enlightenment (*bodi setgel*), illustrating the foundational principles of eternal and complete union of skills and wisdom (*arga bileg*) alongside the complementarity of men and women representing all living beings in all realms at all times. The soyombo symbol has appeared on the national flag, postage stamps, and the coat of arms at different times in Mongolian history. It appears also on the national currency, and thus the *tögrög* is an important mnemonic object for Tömörbaatar, reminding him of Zanabazar's higher wisdom and the "true value of things" (*zöv üne tsen*). To display appropriate reverence, Tömörbaatar has his own distinctive view of how money should be physically handled. He prefers that money notes should not be folded, least of all become dirty or creased. They should be kept in a large, good-quality wallet where the notes should be arranged in a systematic sequential order with all of them facing the same way. Only by physically handling money in this way can people, in his view, ensure that the material side (*matyeriallag tal*) of wealth does not take priority over our intellectual minds (*oyuun uhaan*). It is for him part of an instructive and disciplinary regime that serves to ensure that noble ideals do not become base practices.

However, like most of the other lamas, Tömörbaatar does not see a linkage between physically dirty money and concerns about pollution, as long as money is handled respectfully. So when disputes arise over gold money, it is not because of its polluted status. It is not because lamas fear that the monastery itself is becoming a "magnet of misfortune" (Højer 2012, 38) or that the lay accountant is failing to set aside the gold money for perishable purchases only, as described in the previous chapter with regard to herders' preferred use of polluted money. In contrast to the herders, ninjas, and shopkeepers in Uyanga whom I have described so far, the lamas view the materiality of money in a radically different way. Rather than being a vector of pollution, gold money is a welcome source of wealth that can help sustain the village monastery to an extent that it never has before. But over the years, some lamas have started to raise the issue of money's distribution. Vast quantities of money have been

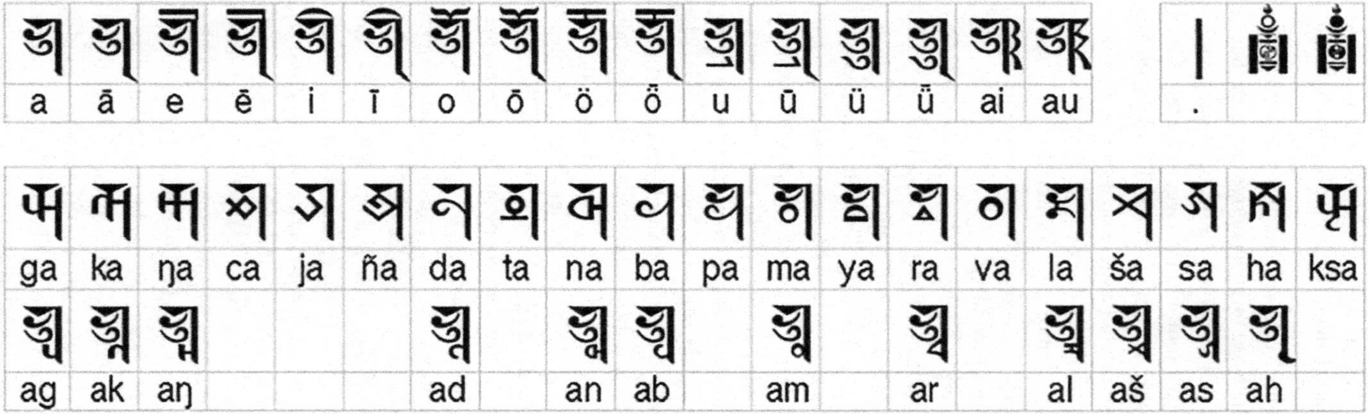

Figure 17. Soyombo script. Font courtesy of Jason Glavy.

Figure 18. Soyombo symbol. Font courtesy of Jason Glavy.

entering the monastery for more than a decade now, and some would like to receive a bit more, especially the younger generation of lamas, who often do not receive generous personal offerings (see also High 2016b). However, so far the elder lamas have refused to engage in any such conversation. For Tögslam, mentioned in the introductory vignette, such complaints are signs that self-centered freedom might be entering even the monastery.

This chapter has shown how the village lamas, as mediators and mitigators in the cosmoeconomy of the gold rush, have become centrally involved in the circulation of gold money. Ninjas and company bosses increasingly rely on the lamas' ritual assistance to ensure profitability and safety in mining, as do relatives of the deceased when such assurance fails. In Uyanga, it is thus the religious specialists, not the ninjas, who are becoming noticeably wealthy because of the mining. But this is not to say that the village lamas are therefore impious, that prosperity and Buddhist practice are necessarily mutually conflicting. Alongside the image of the renouncer, epitomized in the hagiographies of the Buddha, dramatically different conceptions of Buddhist material life also saturate the early Buddhist Pāli Canon and later historical accounts. Rather than challenging their religious insights, I suggest that the lamas' wealth demonstrates the laity's trust and confidence in their abilities. As the lamas welcome opportunities to help the laity, including the red, quarrelsome people of ninjas, and readily accept their gold money, the village monastery has thus become intimately implicated in the gold rush and the negotiation of its potential longevity.

But the lamas are not the only ones in Uyanga who are managing to capitalize on the fortunes of gold. In the final chapter I will show how the local gold traders have become experts at "renewing money" by drawing on an extensive network that intersects with foreign currency flows.

6 Trading Gold

AFTER A LONG stay with an *altny chanj* (unregistered petty gold trader, lit. changer of gold) in Ölt, I returned to Yagaanövgön's household on the steppe when a van suddenly pulled up next to the main *ger*: "Mette! Mette! The police are here! They want to talk to you!" Breaking the afternoon slumber, my host mother's penetrating scream startled and confused me. What did they want? What did they know? Reluctantly I walked across to the main ger, where the police sat waiting by the warm stove—waiting for me and waiting for further servings of strong *shimiin arhi* (home-distilled vodka). I formally greeted the visitors and took a seat some distance away, comforted by the company of my host sisters. The police omitted the customary polite introductions and began questioning me in a hard and intimidating tone: "You were recently staying in Ölt?" "For how long did you stay there?" "With whom did you stay?" "What was the purpose of your stay?" As I was trying to answer their barrage of questions, one of the officers interrupted me and said sternly, "Stop telling us lies! We know what you do. We know that you are an altny chanj. We know that you travel between Ulaanbaatar and Uyanga. We know that you have connections. You buy gold and 'renew money' [*möngiig shinechlegdeh*]! Isn't that what you do?"

That very moment Yagaanövgön entered the ger. One of his sons had fetched him from a distant pasture, where he had been tending a herd of sheep and goats. His arrival brought the interrogation to an abrupt and instant halt. Yagaanövgön took a seat in the honored part of the ger near the family altar. Looking straight at the visitors, he asked, "What are you doing here?" One of the officers started to explain their suspicions of my involvement in the illegal gold trade, but my host father showed limited patience with their cause. Within minutes the police got up and left. As they started the van's engine, one of them shouted out of the window that they would return again soon.

Initially the uncomfortable encounter made me decide to curtail my fieldwork with Uyanga's gold traders. But it also alerted me to practices of money's "renewal" (*shinechlelt*) and "conversion" (*öörchlölt*)—highly specialized practices that are not associated with the village lamas described in the previous chapter. I have shown how the religious specialists are centrally involved in and profiting substantially from the tense cosmoeconomy of the gold rush, which has brought multiple and diverse beings into fraught and untenable proximity. But although the lamas address spirit beings and promote appeasement, they are not deemed capable of renewing money. Instead, this is something that Uyanga's various gold traders are seen to excel at. In this final chapter I trace gold as it leaves the "land of dust" and moves along the paths of the illegal gold trade. Turning to these specialists, I examine their particular abilities to make the fortunes of gold profitable. Hotels, bars and restaurants, auto repair shops, motorbike dealers, and tire vendors, even a wood bucket business are all associated with, if not actually owned by, altny chanj and their so-called *tom darga* (big bosses). After a decade of gold rush mining, these traders have left their own distinctive marks on the village through their seemingly inexhaustible ideas for potential business ventures. Some of these businesses are in operation for only short periods, while others acquire a long and commanding presence. Irrespectively of their specific mercantile foci and longevity, they all exemplify monetary techniques that are central to merchant capitalism (Day 1999)—techniques that have been made familiar in anthropology through the writings of scholars like Eric Wolf (2010) and Marshall Sahlins (1994). As a well-known modality in macroeconomics, the emphatically enterprising fervor of merchant capitalism in places like Uyanga has attained its own term, namely that of "gold rush capitalism" (Eifler 2005; see also Jung 1999). This is a mercenary mode in which people have "learned well the axiom that the main chance for success [lies] not in mining gold but in mining the miners" (Rawls 1999, 7). As in many other historical and contemporary gold rush communities across the world, Uyanga is a place where a defined group of people is setting up businesses that serve the needs and tickle the desires of miners and many others.

But Uyanga's gold traders not only have become prominent and instrumental financiers of mercantile activities. They also have their own elaborate ideas about what role their monetary dealings should play in the cosmoeconomy of the gold rush—ideas that would not be comprehensible if approached exclusively through standard accountancy approaches to wealth creation. As the gold traders seek to optimize their businesses, they regard profit (*ashig*) as not simply dependent on various ways of calculating and influencing the difference between earnings and expenses. As experts in the renewal of money, the gold traders are drawn toward the energy (*enerji*) that they see inherent in the yuan of their Chinese trading partners within the expansive network of Asia's illegal gold trade—an affective quality that highlights money's capacity to become productive and profit-earning capital. It is this *enerji* that enables, if not compels, gold traders to invest in businesses, especially in "creations that can be seen by eyes and touched by hands" (*gart barigdaj nüdend haragdah büteen baiguulalt*). Without the conversion with the foreign currency, the traders deem the proceeds from gold sales unsuitable for the circulation and material objectification of monetary value. As Maurice Bloch and Jonathan Parry (1989) demonstrated long ago, such local understandings of spatio-hierarchical thresholds to exchange do not necessarily reflect a lack of familiarity with or knowledge about trade (see also Gell 1992; Thomas 1991). Nor should they automatically be interpreted as inhibiting barriers, which they might seem at first (see also Munn 1986). In monetary exchange, as

Jane Guyer (2004, 40, 42) notes, asymmetry is in many cases expected, if not actually "cultivated" and "dramatized" by those involved. In contrast to a market ideology that presumes equivalence or indeed Paul Bohannan's (1955) classic model of a monetary market-averse "spheres of exchange," thresholds are here seen by the gold traders to be facilitating rather than crippling, rewarding rather than punishing. Recognizing that "multiple exchange logics are always at work" (Ssorin-Chaikov 2000, 345), far exceeding those approved by formal state regulation and central banks, Uyanga's gold traders thus actively seek positions within international networks of asymmetry. Deliberately creating distance between the calamitous origin of gold and the proceeds from their sales, they increase the liquidity of gold money and its potential for conversion into singular creations. Money from Mongolia's gold rush in this way facilitates the emergence of an elite that is no longer based primarily on the herd sizes and patriarchal prowess described in chapter 1 but rather on new forms of business acumen and high risk taking.

Risky Business

Flowing outward from the mines of Uyanga, the first nodal point in the illegal gold trade is the altny chanj: a gold trader who resides in the mines and buys gold straight from ninjas, as I described in the context of Buyanaa in chapter 4. The gold trader usually operates several businesses that can help entice ninjas to immediately part with some of their earnings. Many gold traders thus operate also a ger shop, a film ger, or a *guanz* (restaurant). Or they might have some pool tables in front of their main ger in case people, for the payment of a small fee, are interested in a game.[1] Of the various altny chanj whom I met over the years, Battsetseg was someone who was very experienced, saw clearly the structure of the trade, and willingly shared her insights.

I met Battsetseg during my very first visit to the mines. At that time I was staying with Nyambuu's family. We drove from their *ail* on the steppe, wound our way through the valley, and continued up a long, steep hillside until we eventually stopped on the other side of the ridge in front of a ger positioned precariously between deep mining holes. When I gingerly stepped out of the car, I noticed a large piece of cloth hanging on the outside wall of the ger (see figure 19). Big handwritten letters announced *alt avna*, meaning that gold was bought at this ger. When we entered, a flustered Battsetseg tried to wake up her drunken husband, Bayasgalan, Nyambuu's oldest son. Bayasgalan was sleeping on the only bed in the ger, and he eventually managed to sit through our visit. For Battsetseg, life as an altny chanj had begun around 2003, and as she later told me, it "changed everything" (*büh yum öörchlögdöh*).

Battsetseg grew up in the same valley as we had just driven through. During the socialist period her parents had worked for Uyanga's herding and milking brigades. Later they became "independent herders" (*höviin malchin*) who sustained a decent-size herd. When gold mining licenses were granted to Erel in the early 1990s, the company activities took place near her home, and for Battsetseg, *tednii büteel* (their doings) became part of ordinary life. Dump trucks and diggers frequently passed their ail, and

Figure 19. Gold trader's *ger* announcing that they "buy gold" and listing "milk, dairy products, meat, medicine, and coal," which they sell

drivers often stopped for a refreshing bowl of *airag* (fermented mare's milk) or a cup of salty milk tea. At some point a gold dredge was positioned in the nearby riverbed to sift through the river gravel. Although her parents were worried about its impact on their drinking water, "the thing" (*yum*) was for Battsetseg a source of immense curiosity, if not excitement. It was big and rusty, noisy and powerful. "Not everyone had such a thing close by. It was interesting," Battsetseg reflected. For her, industrial gold mining was part and parcel of her childhood memories of life on the steppe, and it remained central to how most of her family went on to make a living.

Battsetseg attended the village school, where she met Bayasgalan. In her view, she was much more interested than he in what they taught at school. So while Bayasgalan learned how to drive and became a driver for Erel, she was given her father's permission to go to nursing school in the regional capital of Arvaiheer. Upon her graduation, she returned to Uyanga and Bayasgalan, and she began working at the village hospital.

But her monthly salary was meager, no more than 40,000 MNT (33 USD). Struggling to make ends meet on such a low income, she decided to join Bayasgalan, who had started to wash dirt (*shoroo ugaah*). Apparently, he had had a major argument with his father and had left for the mines in a drunken rage. This argument had been so fierce that people still talked about it years later. By 2003, Battstseg had become so adept at artisanal gold mining that she left her job at the hospital. The couple converted their ger into a small ger shop (*geriin delgüür*). At that time Choidogsüren, the maternal uncle of Bayasgalan, was a tom darga, and he asked them if they wanted to buy gold for him. And they agreed.

Today, Battsetseg and Bayasgalan, aided by their six-year-old son, Baterdene, run a ger shop that sells a wide assortment of goods. Small shelves are packed with everything from shoe polish, superglue, and cigarettes to washing powder, sweets, and alcohol. During the day their bed becomes part of the shop, offering a useful space for cardboard boxes full of goods, and at night the bed once again becomes a family space. Beyond all the shop items, some of the conventional ger furnishings are visible—the stove in the center, the altar in the north, and the bed in the west. But as mentioned in chapter 2, the use of domestic space among ninjas bears little resemblance to that on the steppe. With Bayasgalan spending most days drinking or sitting outside hungover, it is often Battsetseg who serves customers arriving throughout the day. When customers enter the shop, she gets up hesitantly from her stool by the stove and engages with them only when asked about a particular item. She rarely has any direct eye contact with them and no small talk takes place. Battsetseg announces the price, receives the money, and throws it into a cardboard box where she keeps all their change. She then grabs the requested item and passes it to customers along with the eventual change. Transactions are quick and impersonal—a dynamic that I saw repeated across different ger shops in the mines and far beyond.

Bayasgalan and Battsetseg drive to the village of Uyanga about three times a week to buy more stock for their shop. Once in town Battsetseg spends hours going from shop to shop. Rather than buying in bulk, she purposefully spreads her purchases over many shops. Accompanying her on some of these trips, I initially presumed that she did this in order to ensure that her various friends and relatives who were running shops in the village would all benefit from her business. But as discussed in chapter 4, there is also a particular logic to the way in which shopkeepers in the village run *their* businesses. And just like ninja customers, Battsetseg is at *their* mercy.

Village shopkeepers like Pürevtogtoh and Soylham are not keen to receive large amounts of gold money in a single transaction. If a customer spends a lot of money in their shop, they fear it will be too heavy (*heterhii hünd*) and not be passed on so easily. Concerned about money's capacity to "be accompanied by trouble" (*gai daguuldag*), they not only deem gold money less valuable and hence ask for more money but also occasionally outright refuse to sell more goods. One day, just before finishing her purchases at a shop, Battsetseg saw a particular brand of cookies she had not seen before. She told the shopkeeper how many packages she wanted to buy, but the shopkeeper replied, "*Odoo bolno oo* [It is enough now]!" Battsetseg pleaded with the shopkeeper because she knew that she would be able to sell the cookies instantly. "I couldn't buy them," Battsetseg recounted. The shopkeeper had plenty of stock, so she had not refused because the shop was running out of supplies. Rather, as Battsetseg explained in a later conversation, it was because the shopkeeper had enough polluted money. In order for the shopkeeper to still be able to transfer the weight of the money and avoid its undesirable accumulation and stagnation in the shop, Battsetseg had no option but to accept the refusal to sell. Whereas I had presumed that her friends and relatives would be keen to get a large share of her money, it transpired that Battsetseg instead had to carefully negotiate the limits beyond which her money posed too great a threat to others. Although shopkeepers in the mines have more money than most other customers in Uyanga, even they have to also accept the dangers associated with their money. Given the cosmoeconomic expectations in the area, the greater affluence and collective buying interest of Battsetseg and the other gold traders did thus not afford any greater buying power.

Battsetseg was well aware of the village shopkeepers' caution toward her money, and she was not surprised by the reluctance to accept her dirty money notes. "Well, I do it too," she shrugged before adding that it was perhaps not exactly the same. She did not share the village shopkeepers' view that gold money was dangerous, heavy, or polluted. But she did ask for "a little more" (*ahiu*) when people gave her tattered money notes. "Everybody here gives me dirty money notes so I generally just raise the price," she explained. When reselling her purchased goods in her ger shop, she thus applied a substantial profit margin of no less than 20 percent. I estimate that in 2005 the shop had a turnover of around 1,000,000 MNT (833 USD) every month, providing Battsetseg and Bayasgalan with a monthly net profit of around 200,000 MNT (167 USD), much more than she would earn as a village nurse.[2]

But for my hosts, the *ashig* (profit) of the shop was inseparable from their involvement in *ersdeltei business*, a term I translate as "risky business," commonly used to refer to unlicensed gold trading.[3] In Mongolia, the Minerals Laws of 1997 (Mongol Ulsyn Ih Hural 1997) and 2006 (Mongol Ulsyn Ih Hural 2006) stipulate that the Bank of Mongolia (Mongol Bank) is the only legally authorized purchaser of gold in the country. Aligning itself with the internationally recognized benchmark for pricing gold products on the world market, the bank offers the London Gold Fixing 3:00 p.m. price to domestic producers.[4] However, the bank, which is located about five hundred kilometers away in Ulaanbaatar, requires people to undertake a twenty-four-hour journey to sell their gold. Moreover, when selling gold to the bank, one must pay a variety of taxes. At various times during my fieldwork, the royalty tax on gold was between 5 and 7.5 percent of the sales value. In 2006 a windfall profits tax of 68 percent on gold was also introduced, which outraged mining companies because it was the highest such tax in the world at the time (Namjil 2008). It has since been repealed and replaced by a surtax royalty rate of up to 5 percent depending on the purity of the traded gold. Last, gold sellers are also required by law to show evidence of a valid mining license, something that is not possible for artisanal miners in Mongolia to acquire and, as I have discussed elsewhere (High 2012), is a legal option only for mining companies (yet see

SAM 2008 and Mongol Ulsyn Ih Hural 2008). Given these unfavorable conditions for authorized gold trading in Mongolia, it is not surprising that ersdeltei business is rumored to shift as much gold as the formal trade, if not more.

When ninjas visit Bayasgalan and Battsetseg to sell gold, the dynamic in the ger suddenly changes. Bayasgalan or Battsetseg quickly takes a seat at a small table near the altar and arranges the necessary equipment, which is otherwise stored away in a cupboard: a small digital scale, a calculator, a container for the gold, and a black purse. The customer is invited to pass the stove and stand by the trading table. If accompanied by others, however, he or she is instructed to wait quietly by the door until the transaction is completed. The customer carefully presents the gold, which is usually kept tightly wrapped in the protective foil from cigarette packs, and hands it over to a quiet but attentive Bayasgalan or Battsetseg. The gold is slowly poured onto the scale, and if there is less than 0.1 grams, it is rejected as too insignificant an amount for a trade. I estimate that about one-third of ninjas come with such small amounts. As soon as the weight is displayed on the scale, Bayasgalan or Battsetseg pours the gold into the small container, which is put away immediately. Only then do they open the black purse and hand over the corresponding amount of money to the customer. The ninja usually decides to buy something from the ger shop before leaving, be it a bottle of vodka, some cigarettes, or a handful of cookies. The turnover of the gold trade is in this sense directly intertwined with the turnover of the ger shop. A profitable transaction in one business sets in motion a profitable transaction in the other. When I asked Bayasgalan if they ran their businesses as two separate entities, he said, "Do you see this?" He pulled out the black purse. "This is my *nya-bo* [accountant] and this is what the police want." Once, during my stay with them, the police made an unannounced visit. Suddenly we could hear a *furgon* (a Russian minivan) approaching the ger. Bayasgalan quickly finished a gold trade and locked away their equipment, while Battsetseg grabbed an unopened box of vodka, positioning it so it was ready to give to the police. The door flung open and half a dozen policemen stormed in. They searched through the cardboard boxes and aggressively pushed and threatened Bayasgalan and Battsetseg until eventually leaving with the box of vodka and several packs of cigarettes. They did not find the black purse, but Battsetseg knew several cases in which police had taken the money of gold traders.

For Bayasgalan and Battsetseg, the black purse is central to how they run their two businesses. Every evening they transfer money from the cardboard box used for shop dealings to the purse with gold money. When they buy stock for the shop, the money also comes from the purse, as does the money they pay ninjas when buying gold. The black purse thus holds their circulating currency, linking transactions in one business with that of the other. But the purse not only stores their sales revenue. It is also seen to affect the stored money and make it profitable and productive (*ashigtai*) capital. If money is not put into the black purse, they consider it *ühmel möngö*, that is, dead or lifeless money. This money is undesirable and cannot be invested in businesses, lest it become *ühmel höröngö*, which are "dead assets" that have neither present nor prospective value. Among the gold traders, a quintessential image of ühmel höröngö is as an unfinished and abandoned building: its structural incompletion highlights that it failed to become a business, yet it retains a physical presence that reminds the investor and passersby of the investment made. It is a venture that is without vitality and will remain forever unusable.

Among gold traders, the distinction between productive and dead money does not rely on conventional theories in monetary economics. Although they use a national currency and engage in mercantile activities that are common to gold rush communities across the world, the traders view wealth creation as not only a matter of attracting customers, setting the right prices, keeping expenditures low, investing in other businesses, lending money to others, and the like. These strategies are certainly deemed crucial, but without the physical object of the black purse, there would be no profit—a capacity of money that is considered isomorphic with the material artifact of the purse and the renewed money that has been acquired from the big bosses of the illegal gold trade. For them, it is the physical closeness, the actual touching, of the renewed money with their own income that helps engender productive capital.[5] Just like market traders touching the remaining goods with the received money immediately after a transaction (see Højer 2012, 38), physical objects are seen to be able to pass on the fortune or blessing in "the taking and giving" of goods. Empson (2011) has described how more generally physical objects can act as containers or vessels of fortune. If people are separated from these vessels or if the vessels are destroyed, their human holders will experience misfortune (see also Empson 2012). When fearing for the loss of the black purse during police raids, Bayasgalan and Battsetseg are therefore not only concerned about losing a substantial amount of money. They are also concerned about losing their key to profit making.

Renewing Money

In a region where the vertical organization of social relationships is central to the so-called Mongolian episteme (Pedersen 2001, 419), access to renewed money from the big bosses of the illegal gold trade is available only to those few who are positioned favorably within relevant hierarchies.[6] In these particular "circuits of commerce" (Zelizer 2004), which differentiate and regularize ties and transfers, the big bosses are well defined at the apex. They control transactions and set boundaries; they employ distinctive media, in this case renewed money; and they orchestrate an array of organized, differentiated transfers involving international gold buyers in Ulaanbaatar as well as Uyanga-based altny chanj and ninja customers. Reminiscent of "the principle of the single leader" that was promulgated through the collectivization of the economy during the socialist period (Humphrey 1998, 106), these big bosses are men who dominate and issue commands that subordinates can do little about (ibid., 110). Given their "power over people, places and resources" (Zimmermann 2012, 84), they have become instrumental gatekeepers to a kind of money that is much desired locally.[7]

One of these big bosses is Choidogsüren. He is the older brother of Degidsüren, Nyambuu's wife, and he exudes a striking, almost intimidating, confidence. According to Degidsüren, ever since Choidogsüren was a small boy, he displayed not only a passion for learning but also commendable obedience and wit. While he was growing up with his nine siblings, this combination of qualities seemed to help him out of almost any difficult situation. In the eyes of his sister, he was their parents' favorite child, and they would do anything for him. In 1957, they moved to Uyanga so that Choidogsüren could attend the local school. Graduating ten years later with a fine diploma and a praiseworthy record of involvement in socialist youth groups, Choidogsüren was selected to go to university, and he eventually became the chief editor of a well-known newspaper based in Arvaiheer. In 2000, when gold mining in Uyanga burgeoned into a gold rush, Choidogsüren decided to move to Ölt. But rather than joining the ninjas in washing dirt, he positioned himself as one of the first altny chanj. As an already influential person, he drew on his extensive network of contacts, and he knew someone in Ulaanbaatar who would buy his gold. Money soon started flowing in, and he received constant inquiries from friends and relatives who wanted to join his profitable business.

When I met him in 2006, he had become a tom darga. He had positioned several of his friends and relatives, including his nephew Bayasgalan and Battsetseg, as his altny chanj. Since he was no longer buying gold straight from ninjas, he had moved out of the mines and settled with his family in an imposing house in the village. However, he was rarely at home. Most days he was either paying visits to his altny chanj or driving to Ulaanbaatar, where he sold his gold to a Chinese buyer (see figure 20). Unfortunately I could not accompany him on his trips to Ulaanbaatar. Choidogsüren had established a good and stable contact in the capital and did not want the Chinese buyer to grow extra concerned about the business. The illegal trade was already a priority for the police, and Choidogsüren did not need a foreigner like me to increase risks any further. Driving a new Toyota Land Cruiser with black-tinted windows, owning Uyanga's one and only hotel, and running several shops in the area, Choidogsüren, in his own words, was "doing well for himself" (*az hiimortoi baisan*, lit. was with luck and fortune).

In the unlicensed gold trade in Mongolia, three weight measurements are important when calculating the monetary value of gold: the *fun*, the *tsen*, and the *lan*. These measurements correspond roughly to the historical Chinese weight measurements of *fen*, *tsin*, and *tael*, which were used for trading precious metals such as gold and silver. The *fun* is the smallest unit, equaling 0.37 grams. For ninjas, it is generally disappointing if they have only a fun of gold for a trade. As I described in chapter 2, ninjas are expected to share the money they receive for their gold equally with the others on their mining team, and with a fun each person would receive hardly anything. The term *fun* is therefore most commonly used in a derogatory way in banter within mining teams when mocking others for their poor mining skills. A much more desirable and important amount is the tsen, which equals 3.75 grams, or ten fun. When ninjas sell gold, the value is always calculated according to the tsen price: (weight in grams/3.75) * *tsen* price = value MNT.

Figure 20. On their way to the capital

In June 2005, Uyanga's altny chanj paid ninjas 48,000 MNT (44 USD) for one tsen of gold, depending on its purity. Within a year it rose to 66,000 MNT (59 USD). With a year-on-year record-climbing gold price, the tsen price hit 155,000 MNT in April 2010 in the neighboring region of Bayanhongor (Mönh-Erdene 2011, 57).[8] The tsen price may vary slightly between different mining camps, depending not only on the quality and amount of gold but in my view also on the relationships among the tom darga. In Uyanga, at the time of my fieldwork they all knew each other well. According to Choidogsüren, they met regularly to agree on a set tsen price, which followed the fluctuations in the London Gold Fixing. In 2006, there was apparently a new trader who was driving around Ölt, offering a higher tsen price to ninjas. The ninjas caught on, as did the altny chanj. The altny chanj reported their competitor to the police for illegal trading, and he was said to have never returned to Uyanga. Although the altny chanj could hypothetically offer ninjas a marginally higher tsen price and still take a cut for themselves, there is a general adherence to the price fixing agreed by the tom darga in Uyanga. Uniting the altny chanj and the tom darga into a single powerful alliance, the local price fixing immediately communicated to others who was a rogue and who was a recognized participant in the risky business of Uyanga.

According to Choidogsüren, each of his altny chanj purchased approximately 40 grams of gold within a typical twenty-four-hour period during the contracted peak mining season. Depending on the amount of gold sold to the altny chanj, Choidogsüren visited his petty traders a couple of times a week. Since so much gold was being traded, it was talked about in terms not of tsen but rather of lan. This is a weight measurement of 37.2 grams, or around ten tsen. During my stay with Bayasgalan and Battsetseg, Choidogsüren paid them an additional 5,000–8,000 MNT per lan of gold, depending on its purity. Although the profit margin for the couple might seem very small, between 0.5% and 1%, they traded so much gold that the sales still brought them a lucrative income. With the flexible and locally negotiated value of money, their income was at least about 200 USD per month in addition to the income they received from the ger shop.[9] Yet, as Battsetseg noted, profit "glues" (*naaldah*) to the tom darga, not the altny chanj. Choidogsüren would not disclose how much he received for his gold when reselling it to his Chinese partner, but there was a price difference of 150 USD per lan between the London Gold Fixing and the price he offered his altny chanj. Compared with figures reported for the neighboring region of Bayanhongor a few years later (Munh-Erdene 2011, 95), the profits taken in by the tom darga in Uyanga seem to have been much larger than those earned elsewhere. This is perhaps partly due to the alliance between the tom darga and their ability to price-fix gold. That Choidogsüren also drew heavily on his personal networks and used predominantly junior friends and relatives as his altny chanj probably further facilitated this compliance. Rather than being confronted with discontent and requests for higher monetary returns, Choidogsüren seemed surrounded by people who were keen to remain his gold buyers.

Choidogsüren liked to brag about his various achievements, especially his business ventures in Uyanga. For him, this success had been possible partly thanks to his long-standing Chinese business partner in Ulaanbaatar. In 2006, when the international gold price spiked past 700 USD per ounce, the Ministry of Mining together with the police and the Bank of Mongolia began making public announcements on daily television, condemning the illegal trading and exportation of gold. People were urged to pass on any details about the trade and were promised legal protection as well as substantial monetary rewards pending the quality of the information revealed. Faced with public campaigning and heightened attention, Choidogsüren depended on a trustworthy gateway to the global market in illegal gold to trade his many lan of gold. He paid an unknown premium for this access, but he said that there were plenty of traders from various countries who could facilitate the trade. However, Choidogsüren wanted to trade only with Chinese people (*Hyatad hümüüs*). In a country where scholars have documented much general hostility toward the Chinese (see, for example, Billé 2013, 2014; Reeves 2013), this desire initially surprised me. But I realized that it had little to do with Chinese people and much more to do with their Chinese money.

One day, after a successful trade in Ulaanbaatar, Choidogsüren returned home for a lavish celebratory meal of mutton dumplings (*buuz*) and plenty of vodka. Drying the grease off his hands, he leaned back, sighed deeply and said,

> People think that we deal with gold. That we destroy nature. But we really deal with money all the time. We are not gold changers. You know them. . . . They are all in the mines. We are money changers [*möngönii chanj*]. It's a very risky job because the chance of getting connected to dead money accidentally is high [*ühmel möngötei sanamsargüi baidlaar holbogdoh bolomj öndörtei*]. What matters is money's channel [*suvag*], money's origin [*üüsver*]. I make the origin good. I give our Uyanga good money.

"How do you do that?" I asked. "How do you make the origin good?"

> I renew money. . . . Why do you think I go to the city? Why do I travel the long way so often? Money has energy and this energy depends on money's origin. So money made from bad karma acts have bad energy. Money made from theft, money made by lying. This kind of money has bad energy and it's now all over Mongolia! But there is also money that has good energy and every time I go to the city, clean good money comes to me [*nadad ireh*] as something precious and valuable.

I nodded and remarked that a lot of tögrög was coming his way.

> No! Of course it isn't tögrögs! [He laughed.] It's just like the big mining companies—huge amounts of money coming in, all foreign investment [*gadaadyn höröngö oruulalt*]. Money from outside! The companies that are doing well in Mongolia, the companies that are profitable, where do you think their investments come from? Not Mongolia! It isn't tögrög. It comes from China! It is *Yuan—Hyatad ulsyn möngön temdegt* [the national currency of China]. Companies with Chinese money all become profitable.

Trading his gold for yuan currency was for him so valuable that he admitted to taking a lower price for his gold from his Chinese partner on the condition that he was paid in Chinese money. He said that he could get a higher price from other traders and also avoid a loss of money if he did not have to rely on unfavorable exchange rates. But he *wanted* the foreign currency. I mentioned to him that selling his gold for foreign currency probably made his earnings less traceable for the police, to which he agreed. Yet he noted that the police were not really a problem since they "always like to eat," implying that they were willing to accept payment if they caught him in return for letting him go (Sneath 2006; Zimmermann 2011). For him, a much more important issue was the potential of profitability that he associated with the yuan and deemed absent in the tögrög. When approaching the value of money, Choidogsüren did not consider simply the number imprinted on the money note or minted on the coin. He also actively drew on and sought an exchange that was premised on asymmetry and evidenced in the unmatched economic growth of the Chinese economy. Value thus exceeded the numbers and included qualitative opportunities for gain that lay in thresholds of conversion (cf. Guyer 2004). Rather than selling Uyanga's gold *within* national currency regimes, he strove to create profitability out of a lower face value. Whereas to an outside observer the thresholds of exchange might seem adverse and undesirable, for the tom darga they offered desirable opportunities for profit making.

A New Elite

As a result of his successful renewal of money in Uyanga, Choidogsüren impressed upon me that he felt his life was now in constant danger.[10] He felt that he could not leave his compound unaccompanied or take part in drinking binges with his friends. He sensed dangers lurking everywhere. The dangers were not associated with the police, the company security guards, or the road bandits. Instead, as a noticeable and influential person in the transformation of precious metal into money, Choidogsüren felt that thousands of drunken and often aggressive ninjas followed his every move. Holding and handling unmatched amounts of money captured people's attention, and the risk of a personal attack seemed imminent. As a result, money had to circulate, or as he put it, "It is only when money isn't moving that it is dangerous" (*möngö oruulahgüi bol ayuultai*). Receiving so much money from his trades, Choidogsüren felt forced to quickly convert his earnings into a less extractable, nonmonetary form. Although he could have offered his altny chanj higher tsen prices for their gold, he opted instead for buying luxury cars, giving generous gifts, and investing in businesses. Sharing in the same predicament, all the tom darga have now come to dominate the village through their multiple and diverse ventures.

The tom darga are well known in the area, also among people who are not directly involved in the gold rush. As my fieldwork took me from household to household across Uyanga, people often shared their opinions about the traders, whether or not they had ever met them personally. Rumors circulated about how Uyanga's gold went to China, and many residents were not as thrilled about the Chinese money as Choidogsüren was. Whereas Choidogsüren felt that he was bringing good money back to Uyanga, some felt that it was rubbish (*hog*). In their view, if money was made from gold mining, it would always have black footsteps and bad things would inevitably happen. It was merely a question of time. And as more and more new businesses cropped up across Uyanga, the physical manifestation of the extracted gold became ever more apparent and ominous. Others observed that the tom darga had become exceptionally wealthy. In an area where many herders, ninjas, and villagers were struggling to make ends meet, they wished the traders would be more charitable. They commented that the tom darga seemed to be afflicted by the greed associated with gold, and opening new businesses became an indication of this. Wanting to extract more wealth from others, businesses indicated the traders' parasitical relationship with others, emblematic of what some Mongolians have termed "wild capitalism" (*zerleg kapitalizm*; see Empson 2011, 308). Although Choidogsüren enjoyed much admiration within his network of gold traders, it was far from equally shared outside.

The frustrations and disapprovals of the tom darga, however, were dwarfed by the rapidly growing personal networks of the traders. In addition to the altny chanj, the tom darga had numerous other people working directly for them in their various businesses. Their personal networks were extensive and ever-growing, involving people throughout the region. Moreover, with every new business venture, another distinctive mark was left on the village. A new service, a new product, if not a new building.

When I returned to Uyanga in 2011, my host sisters excitedly showed me the new supermarket and then the new Internet café. These were yet other enterprises owned by the tom darga. With the price of gold at a record high, the tom darga had become a defined group of not only keen but also highly capable investors. They had come to constitute visible nodal points that were brimming with money, whether considered polluted, profitable, or otherwise.

This concentration of wealth among the gold traders has in important ways altered the local "topography of wealth" (Ferguson 1992). Although it was not a radically new elite that had emerged, the kind of wealth on which it based its position was new.[11] Rather than drawing on the herd sizes and patriarchal prowess described in chapter 1, the traders have achieved a position of influence and importance through their business acumen and willingness to take great risks. And instead of investing in pastoral wealth like the herding patriarchs, the traders pursued mercantile investments as their favored strategy for wealth creation. In this chapter I have shown that the risky business is risky not only because it deals in illegal gold but also because it depends on practices of renewal and conversion. That is, it hinges on international trade relationships that can bring in the much-wanted Chinese currency. Receiving Chinese yuan is deemed fundamental to every aspect of Uyanga's illegal gold trade, whether it is the altny chanj running multiple businesses in the mines, the tom darga selling gold in Ulaanbaatar, or the entrepreneurs setting up new ventures in the village and beyond. Relying on the enerji associated with Chinese currency, gold traders do not consider money's liquidity an inherent quality governed by money's material form. Instead we have seen how it has to be actively created through practices of exchange. While this dependence on exchanges involving yuan makes the traders particularly vulnerable to unfavorable deals in Ulaanbaatar, it also makes them particularly successful in Uyanga when it does work. Emerging from and reasserting distinctions between productive and lifeless money, they have transformed their earnings into an emphatically profit-generating artifact unlike money held by most others in the region.

Having become a feature of local life, the gold rush trade has created an economic landscape that resembles many other gold rush communities across the world. With its flourishing mercantile fervor reinforcing divisions between miners and traders, Uyanga's gold rush trade offers a poignant example of a contemporary form of "gold rush capitalism" (Eifler 2005; Jung 1999). Thus it might be tempting to dismiss the traders' own concerns about money's energy and consider it unimportant, if not irrelevant, to the actual workings of the trade. Given that the notion of enerji results in a familiar macroeconomic modality, why reflect in this chapter on the capacities that people themselves ascribe to money? Apart from the inherent dangers of reductionism in any model of abstraction, it is important not to confuse emerging similarities with causal mechanisms (see also Maurer 2008). Rather than presuming that there is a universal logic at operation in gold rush economies that disciplines the various participants into conformity, we see that local practices and motivations inform the ways in which familiar divisions between miners and traders have emerged in Uyanga. For the new elite, their risky business depends on the negotiation of dangers, demands, and differences that they themselves see within the expansive network of which they are a part. It is by recognizing these processes of negotiation that we can understand the nature of Mongolia's illegal gold trade and its strikingly uneven distribution of wealth.

This book began with an observation: in Mongolia's gold rush, people handle gold money as a controversial object of potential misfortune. Thousands of people have sought to go to the mines in search of the precious metal and have been remarkably successful. Large amounts of gold are bought and sold on a daily basis, which sets in motion an intense circulation of money. But the gold money is subject to multiple and diverse conceptualizations that affect its subsequent circulation. Understanding this dynamic is crucial for understanding not only the phenomenon of the gold rush but also the object of money.

This book set out to explore this dynamic by analyzing how herders, ninjas, local shopkeepers, Buddhist lamas, and gold traders take part in the emerging gold rush economy. Directly affected by changes in the international metal markets and currency exchange rates, these gold rush participants are intimately familiar with the workings of money. When money associated with gold mining is handled in conflicting ways, it is thus not an expression of people's lack of familiarity with state currencies, nor is it a critique of bewildering currency regimes. Their views of gold money are not grounded in ignorance, misapprehension, or unfamiliarity. Instead they are informed by extensive and direct participation in and contemplation of the workings of money. Just like any other object, gold money is a social medium that is inextricably part of its wider world, and this is something that people in Uyanga are intimately aware of. I have shown that rather than being exclusively authored and authorized by a state and its central bank, the tögrög has slipped into alternative value regimes that involve both humans and nonhumans. It is subject to interpretation and contestation, representation and resignification. Spirit worlds with powerful spirit beings, invisible dust, and malevolent ritual practices inform how people hold and handle gold money, as do notions of transferable pollution, impending misfortune, and money's renewal. This symbol of value equivalence is far from the alienable, anonymous, and nonlocal medium that is the national, and indeed macroeconomic, premise of a currency. As we have learned here, the ways in which people make sense of their money inform and directly affect how they decide to use it.

Money's capacity for symbolic representation has been richly documented. This "symbol of all symbols" (Gregory 1997, 33) lends itself well to expressing teleologies, whether of hope or despair (Maurer 2006). It has provided a conceptual basis for a variety of influential analytical frameworks that seek to understand how people make sense of change (e.g., Bohannan 1959; Taussig 1980; Bloch and Parry 1989; Robbins and Akin 1999; Comaroff and Comaroff 1999). It can offer an idiom, a symbolic language, a metaphor, if not a memory (Cribb 2005; Crump 2011; Hart 2001). Evidencing the incredible potential of money for symbolic representation, these studies have detailed how money can readily become part of discursive spheres of sense making. I have shown that in Mongolia this is no different. Gold money has become an index of morality, communicating people's understandings of the precious metal, the act of mining, and, more broadly, the high stakes involved in the cosmoeconomy of the gold rush. But beyond the discursive sphere, gold money has also become a material vector of pollution, physically carrying and transferring capacities of misfortune as it circulates through the region of Uyanga. And furthermore, I have argued that these representations also apply to and affect its use as a means of exchange, a unit of account, a means of payment, and a store of wealth—that is, the definitional core of what makes money a currency. Rather than assuming that central banks are exclusively in charge of the symbolic attribution of value in money, I have demonstrated that in Uyanga it is also the people themselves who evaluate and decide the exchange value of the legal tender. The people who utilize the state currency draw on and create not only a powerful figurative language but also an altered fiscal reality.

In order to place people's own conceptualizations of gold money at the heart of my analysis, I have used the notion of cosmoeconomy. By doing so, I have not intended to infuse economic life with the traditional static holism portrayed in earlier studies of cosmology (Abramson and Holbraad 2012). Nor have I wanted to assume that economic life is necessarily one of consensus or harmony. Indeed, as fieldwork took me from household to household across Uyanga, my hosts were often keen to point out that economic life entailed the participation of both human and nonhuman forces. Since these participants did not always share the same predicament, the interface of cosmology and economic life was an unavoidable and active challenge for my hosts. This tense and often fearful symbiosis did not simply emerge periodically in moments of reflection but constituted the very condition for people's actions. Without discrediting or belittling the significance of this symbiosis in Uyanga, we should not presume that such challenges are unique to the Mongolian gold rush. Indeed, given the wider importance and recognition of spirit worlds in the Mongolian cultural region, it is hard to imagine why such challenges of coexistence would not also occur in other trades and industries. Regional scholars have helped us understand intricate dynamics in nomadic pastoralism, but there are so many other, often interrelated, livelihood practices that are still awaiting scholarly attention. Rather than expecting these challenges to be the same, or even similar, this is an invitation to understand cosmoeconomic life in its multiplicity. And beyond the Mongolian cultural region, the notion of cosmoeconomy might also help us see the actions, interactions, and transactions involving various beings where we might expect to see only humans. As Bruno Latour (2009) has shown, the difficult, if not troublesome, reality of coexistence is also evident in contemporary Western concerns about the global technosciences. Styrofoam cups, oil tankers, and genetically modified organisms (GMO) now populate the earth, having acquired an existence of their own with unknown capacities to influence our lives. GMOs, for example, do not behave in controllable or even predictable ways. Other species that they come into contact with are not in a position to choose whether or not to become implicated in the evolutionary trajectory of a new, scientifically generated species. Since so many of these entities are simultaneously toxic and attractive to one another, Kerry Whiteside (2006, 105–6) aptly asks, "Can we live together?—where 'we' is people and all the nonhuman phenomena with which they become entangled." These symbiotic entanglements are not necessarily mutually beneficial. They do not entail a peace proposal (Latour 2004, 450). Instead they highlight that "living together" is, as Gabriel Tarde has noted, an association of passionate interests where conflict is inevitable (Latour and Lépinay 2009, 39).

Recognizing this fundamental coexistence of conflicting interests in the gold rush, we must ask how commonality is constructed without implicitly valorizing collectivity. This is particularly important when analyzing the object of money—a key symbol that is premised on some degree of shared value. In order to pursue this aim, this book has not drawn on the extensive literature on moral economies. Instead I heed the warnings of Bill Maurer (2009, 258) when he says that "we should be wary of the juxtaposition of morality and economy itself and . . . we should allow it to continue to trouble us." This caution against juxtaposing morality and economy is particularly important given the kind of value judgments that this act of association often entails. Rather than proposing conceptions of morality and economy as open, uncertain, and troubled, the notion of moral economy is often premised on a conviction of which particular economic practices qualify as "moral." And this conviction is not simply a critical response to neoliberalism's "new spirit" (see Zamora and Behrent 2016). It also evinces the continuing influence of intellectual traditions cemented by classical social theorists such as Karl Marx, Émile Durkheim, Max Weber, and many others.

A recurrent distinction in these classical works is a conceptualization of economic value (often phrased in the singular) as distinct from, but interrelated with, moral values (often phrased in the plural). For Marx (1932), this distinction was useful because it facilitated an understanding of how economic relations affected morality, or what he called the collective flourishing of the human "species-essence." Under capitalist modes of production, no such collective morality was deemed possible, and it was necessarily divided along class interests until a new kind of economic structure was brought about. For Durkheim (1893), this distinction between value and values allowed him to study the relationship between the division of labor and the establishment or breakdown of moral orders in society. If labor relations evolved too rapidly, people's moral ideas could lack momentum and cause "social anomie." And for Weber (1904), it allowed him to study the influence of religious ideas in the emergence of modern capitalism. The Protestant ethic encouraged men to apply themselves rationally to the occupational tasks they were called to carry out. The specific dynamics of one category thus brought insights into the specific dynamics of the other. While helpful in the advancement of classical social thought, this dichotomy continues to tear apart, distill, and essentialize conjoined imaginings of economy and morality (see also Eiss and Pedersen 2002; Graeber 2013). Viviana Zelizer (2004, 123) has pointed out that "since the nineteenth century, social analysts have repeatedly assumed that the social world organizes around competing, incompatible principles: gemeinschaft and gesellschaft, ascription and achievement, sentiment and rationality, solidarity and self-interest." In this "hostile worlds" view, which involves opposed paths for societal transformations, favorable moral action has been clearly positioned among that which seeks to further solidarity, community, and collectivity as opposed to individuals' self-interest.

The foundational works that popularized the notion of moral economy focused on noncapitalist sentiments and practices, whether among the English working classes (Thompson 1971) or peasants of Southeast Asia (Scott 1977). They focused on peasants' riots against the ruling classes to ensure fair rather than free-market prices. Strategies to minimize economic risk rather than to maximize potentials for profit were elevated to cross-cultural models for morally sound economic behavior. These moral economies were drawn from networks of mutual support and assistance that ultimately sought to bring about or ensure the collective good. And the participants, at least in these foundational works, were exclusively human. No other beings were afforded the capacity to act as agents and be involved in the definition of this "collective good." While these works highlight the important limits to capitalism's hegemony and humans' great potential for collective action, their identification of what is moral thus remains unelaborated. It appears that economic practices are deemed moral simply because they strengthen a collectivity (Laidlaw 2014, 21). Perceptibly observed by Jane Guyer (2004, 17), this valorization of a stable human community with key relations of sharing is not an analytical position but itself a moral judgment (see, for example, Gudeman 2001, 163; Hann 2010, 196). And it is a moral judgment that makes local practices of profit making and the involvement of diverse beings, so central to a phenomenon like the Mongolian gold rush, difficult to analyze (see also Ortiz 2013, 77). Constituting an approving designation for only certain economic practices, it presupposes a moral domain that is not only singular and predictive but also a necessary aspiration shared by all its participants.

This book has explored a phenomenon that has little in common with this vision of a moral economy. In a place where life has become chaotic (*zambaraagüi*) and people have to carefully negotiate their fortunes, money travels through the region as a potent and controversial object of potential calamity. But rather than giving rise to local money-averse spheres of exchange or techniques of bartering based on gold, there has been an intensification in the circulation of money. Although gold money is considered an unstable object, shopkeepers still seek to increase their turnover, as do the Buddhist lamas and the gold traders. Uyanga has become a region not of austerity and restraint but of spending and splurging. And this became particularly evident during my last visit to Uyanga in 2011.

There was mayhem in the village on the first of every month throughout that year. The government had decided to give unconditional cash handouts to all Mongolian citizens. This came about because, while the country's average salary had surged in recent years, about a third of the population continued to live below the national poverty line. The mining boom had set in motion not only a sharp increase in foreign direct investment but also a rise in inequality, inflation, and general prices. Economists were concerned that the boom had caused the national economy to overheat, putting its own citizens at a clear disadvantage (Hook 2012b; Isakova, Plekhanov, and Zettelmeyer 2012). In an attempt to dampen the growing public discontent and avoid the "resource curse" that has afflicted so many other resource-rich countries (Sachs and Warner 2001; Auty 2001), the government decided to make the issue of economic inequality one of its top priorities. It set up Hünii Högjiliin San (the Human Development Fund) with the aim of distributing money from the mining boom to all citizens. The initial contribution to the fund came from a negotiated prepayment from the Oyu Tolgoi copper and gold mine in South Gobi. In February 2010 every Mongolian citizen received a check for 70,000 MNT (58 USD). Within the same year another 50,000 MNT (42 USD) followed before the amount settled at 21,000 MNT (18 USD) per month. These cash transfers were envisaged to help alleviate people's problems with lack of access to education and credit. In Uyanga, however, many people viewed the money in a rather different way.

Long lines formed as people waited to receive their checks. Then other lines formed as people sought to spend them. Rather than spending the money on education or saving it up for a larger purchase, they seemed to prefer to spend it on alcohol. With the cheapest bottle of vodka on sale for less a single US dollar, village life came to a grinding halt as most men joined the village-wide drinking binge. I was staying with Budlam at the time, and with lines also forming in front of his house, I learned that many villagers considered the cash handout "like gold money," accompanied by trouble and misfortune. After all, they reasoned, it was money that came from mining. While international investors hoped that the cash handouts would reduce "resource nationalism," some of my hosts in Uyanga feared that new forms of pollution and angered spirits accompanied the money.

Ties between resources, people, and places continue to inform the unfolding of Mongolia's mining boom as it goes through its various phases. Rather than entailing unidirectional transitions favoring the commodification of resources or the secularization of spirit worlds, people in Uyanga encourage us to think of different ways in which we can conceptualize value and the world we live in. Living together is a challenge, but it is not a challenge we should try to overcome. Rather, it is a challenge that we need to understand and then live within.

NOTES

Introduction

1. Out of the 3,865 licenses, more than two-thirds have been transferred to foreign investors. Nongovernmental holdings are concentrated in seven companies, which control 50 percent of the nongovernmental-owned licensed areas. In 2012, Chinese companies controlled up to 70 percent of resource assets. According to Hook (2012a), "although most projects are small scale they constitute the bulk of current mining activity." Not surprisingly, a fervent form of resource nationalism has emerged in recent years in Mongolia, with much public attention focusing on the development of two major mines located near the border with China.
2. This was a statement made by Robert Friedland, also known as "Toxic Bob," chairman of one of the world's largest copper and gold mines in Mongolia (Oyu Tolgoi) when he addressed an audience of potential investors in Florida in 2005. He further elaborated, "The nice thing about the Gobi is . . . there are no people in the way, there are no houses in the way and there are no NGOs. You've got lots of room for waste dumps without disrupting the populations." His rapacious statements were soon translated into Mongolian and created an outcry among the Mongolian population, who burned effigies of Friedland in protest in April 2006.
3. The enthusiasm for textbook free-market theory is reflected in the fact that in the early 1990s politicians discussed seriously the possibility of erecting a statue of Milton Friedman in central Ulaanbaatar, poignantly at the site of a former Stalin memorial (Tomlinson 1998).
4. The notion of wealth-driven gold rushes is not usually applied to the motivations of miners but often to the motivations of investors rushing to purchase gold bullion in times of financial instability or gold-linked dividends paid by mining companies in times of high gold prices.
5. The Qing government designated Mongolian places known to contain gold as "restricted areas," a term reserved for territories of unique imperial importance, such as imperial hunting grounds and lands rich with sable on the border with Russia. But whereas restricted hunting grounds were designed, in part, to ensure a plentiful catch for state-organized hunts, restricted gold fields were never meant to be productive. When trespassing or mining occurred, policing was stepped up. Local officials justified these policing efforts in terms of an imperial responsibility to ensure that the land remained pure (see High and Schlesinger 2010).
6. See also Delaplace (2010) on Chinese ghosts haunting present-day Ulaanbaatar. The apparitions are identified as souls of Chinese merchants from the Qing period who have come back to search for the wealth, especially gold, that they have left behind in Mongolia. Although most of their stone palaces have been destroyed since the early twentieth century, their haunting presence seems to still be felt.
7. The British geologist Edwin Mills (1929) makes a similar observation: "The Mongols take no active part in working the gold placers, but spend their time with their herds of cattle, troops of ponies, and flocks of sheep roving. . . . They are given employment at the gold placers as guards, watchmen and overseers. The Russian and Buriats have gone in extensively for farming and are also raising stock of various kinds, while the Chinese are satisfied to confine their activities to farming and gold mining" (quoted in Grayson 2007, 2).
8. Archival records mention that after the opening of a mine in Cecen Khan aimag, the "Mongols were all unhappy about serving" in the mines (High and Schlesinger 2010, 302n54).

9. The official records do not include minerals such as gold and uranium. They were clouded by secrecy and protected by harsh punishment if revealed. A secret uranium mine in the north of the country existed with its own town and railway lines directly to the Soviet Union, yet was not on any maps (Bulag 1998, 22–23). The place has given rise to many mythic tales often referred to when talking about the unofficial workings of the Soviet Union and its relationship to Mongolia.
10. My transliterations are based on spoken Mongolian, in which sentence subjects and suffixes such as those indicating the accusative case (*–g*) are often dropped. In Uyanga there is a sense of pride in talking like an *Uyangynhan* (a person of Uyanga), which involves using certain nouns and expressions, in addition to the dropping of sentence subjects and certain grammatical cases. Given the local aesthetics of speech, in which people often contrast themselves with the city dwellers of Ulaanbaatar, I have thus transliterated statements according to local ways of talking.
11. See, for example, the various ways in which Karl Marx, Georg Simmel, Max Weber, Talcott Parsons, and others regard money as inherently transformative of social relations and institutions.
12. While there are several competing Western ideologies that revolve around this barter model of value, choice theory seems a particularly apt reference. Friedman (1962), for example, sees exchange ratio as the equation between the utility of one good and that of another. To him, money optimizes the functioning of the market because people with different goods then contribute to the satisfaction of their aggregate needs. In contrast, Marx ([1867], 1967, chap.1), for example, sees exchange ratio as the equation between the commodities under exchange and the abstract labor required to produce them, with money functioning as a veil that hides the labor involved (see also Elson 1979; Sraffa 1960).
13. Transforming a national currency into distinguishable qualitative standards is not unique to the Mongolian gold rush. The sociologist Viviana Zelizer (1997) has shown how people in the US between 1870 and 1930 actively applied distinctions to money, which affected the ways in which it was used.
14. While regional specialists have demonstrated the sense of incongruence in cosmological understandings, they offer highly diverse accounts of the extent to which congruence remains an ontological possibility. For Pedersen (2011, 35), shamanism is "an ontology of transition" that with the advent of postsocialism has acquired analogies in the topography of the landscape and people's "layered minds" (120). Any notion of congruence thus appears a nostalgic longing for past ways and a present, if not also future, impossibility (see also Pedersen and Højer 2008). For Buyandelger (2013, 266), however, "shamans convey what is known and what can be known," thus offering people a "useful tool" (27) for healing as it brings back and completes memories of the past through forgotten spirits. Congruence is thus a possibility that is central to the shamans' ability to attract new clients as ever-more rituals will disclose the historical relationships between spirits.

Chapter 1. The Burden of Patriarchy

1. Everybody in the area, including his own family, knew and referred to him as Yagaanövgön. This name means "pink elder," and there was no consensus as to when and how he acquired this humorous nickname.
2. In Mongolia most of the hard-rock gold deposits are found in quartz veins, amenable to surface mining or trenching. Using sledgehammers, chisels, and at times explosives, miners remove the gold-bearing quartz vein and smash it into smaller lumps. The crushed ore is then milled, and mercury is often added in order to create a gold-mercury amalgam. When the amalgam is heated, it separates and the gold is finally recovered.
3. Hydraulic mining was first used by Edward Matteson in 1853 during the California gold rush (Rohe 1985). This mining technique came about when the miners realized that if they could process more gravel, they were likely to find more gold. In the California gold rush, hydraulic mining became the most environmentally devastating form of placer mining. Sediment was deposited in the rivers and plains, giving rise to frequent flooding in the Sacramento Valley. Ultimately in 1882, a farmer brought a court case against the region's largest mining company for having destroyed his farmland by dumping tailings into the Yuba River. He won with the ruling that hydraulic mining was "a public and private nuisance," putting an abrupt end to such mining in the Gold Country (see http://www.sierranevadavirtualmuseum.com).
4. Several gold placers in Mongolia also contain platinum, and these are expected to become major exploration targets for both mining companies and informal miners in the future.
5. The herder is here referring to different regions (*aimag*) in Mongolia, several of them many days' drive from Uyanga.

6. When the gold rush broke out in Uyanga, Erel was officially the largest placer gold mining company in terms of its gold production output in Mongolia. In 1999 the company produced 2,340 kilograms of gold, or 24 percent of the national production (Grayson and Tumenbayar 2005, 5). The Erel mining company is part of the Erel Group, which includes banks, insurance companies, cement plants, sewing factories, housing material factories, a private hospital, and a private school.
7. In the following chapter the ninjas Tsegii and Davaa describe a clash between themselves and the police.
8. Using the classical Mongolian script, Vreeland ([1954] 1962, 64) transliterates the proverb as "Usa tsanai oiriin sain, Uraga elegenei holoon sain."
9. If the wife's parents are much wealthier than the husband's, uxorilocal marriage can be practiced (*hürgen oroh*, to join a family as a son-in-law). This is most likely to be the case if the wife's parents have no biological (*törsön*) or adopted (*örgömöl*) sons. However, there is much local stigma surrounding such practices, and it appears rather humiliating for the husband to accept his incorporation into his father-in-law's ail (see also Humphrey 1978, 99).
10. I do not intend to imply that such kinship practices are necessarily related. Indeed, regional studies (in particular the Environmental and Cultural Conservation in Inner Asia Project) show a great degree of variation in how such practices interrelate.
11. According to Atwood, hypogamy was still found among Buryats in the nineteenth century (Atwood 2004, 314). However, with the historically growing bilateralization of kin ties in Mongolia (see in particular Szynkiewicz 1977), such marriage structures seem to be no longer preferred.
12. Just like Bayagalan's wife, Ahaa's wife was always called only Ber. Both of their personal names were known but not used in daily life.
13. The headmaster expected substantial material and/or monetary donations in order to accept a child in the school, and the teachers often charged parents a monthly "fee" per subject that their children were studying.

Chapter 2. The Power of Gold

1. The verb *tatagdah* contains the verb stem *tat-*, the connective vowel *-a-* and the passive suffix *-gd-*. It can be translated as "to be pulled," "to be attracted," "to feel drawn to somebody or something" (Bawden 1997, 335). When talking about altny chadvar, people also used the verb *örtöh*, which carries meanings such as "to undergo," "to be subject to," "to be struck by," "to be affected by" (281).
2. Rather than using the common noun *alt* (gold) when referring to gold, people in the mines usually called it shar yum (yellow stuff, yellow thing). Since spoken words were likely to attract the attention of spirit beings, people feared that if they talked explicitly about gold, these beings would notice human actions and mete out appropriate punishment. In the following chapter I will discuss some of these spirit-related dangers involved in the extraction of gold.
3. There were of course many possible reasons why members of parliament were reluctant to mention ninja miners. Besides recognizing that doing so offered no potential for personal gains, national politicians were probably also aware that if they acknowledged the illegal activity of ninja mining, they would be expected to do something about it. Since there was much uncertainty and disagreement about what could and should be done, it probably seemed a better political strategy to ignore the issue for as long as possible.
4. Erdenebat has changed the name of his party several times. When it was founded in December 1998, it was thus called the New Mongolian Socialist Democratic Party. In April 2000 the name was changed to the Motherland-New Mongolian Socialist Democratic Party, and since January 2005 it has been called the Motherland Party.
5. For a damaging exposition in a nationwide newspaper of the minister of nature and environment's involvement in the environmental degradation in Uyanga, see *Ardyn Erh* (2006a).
6. For the full award video, see http://www.goldmanprize.org/2007/asia.

Chapter 3. Angered Spirits

1. The verb *uurlah* can be translated into English as either "to *be* angry" or "to *get* angry" (Bawden 1997, 392), depending on the context. Such anger can be seen as a positive or justified anger.

2. *Bag* corresponds to county, *sum* to district, and *aimag* to region. In correlating the classification of administrative units of government to spiritual entities, there is a striking parallel to the Andean ethnography by Sallnow (1989).
3. The word *am'tai* consists of the noun *am'* (life) and the comitative suffix of *-tai* (with), hence "with life." It is commonly translated into English as "animate," "living," or "alive" (Bawden 1997, 20).
4. In rare cases I heard of people who could generally see spirit beings. Interestingly, this was not something desirable. Indeed, people with this ability had a reputation for transgressing taboos and carrying out bad karma acts.
5. According to Haslund (1949, 143), "The steppe cattle . . . were owned by the White Old Man of the Steppe, who generously allowed the Mongols to administer and exploit his wealth. But the White Old Man of the Steppe was weak in the face of many temptations. He often gambled with the Dragon Prince [presumably the lus], the ruler of all human sicknesses and lord of life and death, and as the two mighty beings had to pay in their currencies, the poor Mongols always had to suffer for it." In Uyanga I did not hear this or similar descriptions of the White Old Man. Yet many herders have carved figurines of the White Old Man on their household altar. Prior to and during calving season they often place offerings to him on their altar and/or pay Buddhist lamas to carry out appropriate readings. Although there are no canonical texts about the White Old Man (Heissig 1980, 78) in Tibetan Buddhism, he is a central figure in the Buddhist *tsam* masked dances.
6. The word "yellow" is commonly used throughout Mongolia as an adjective to mark something as Buddhist. A *shar hün* thus means a Buddhist lama. However, in Uyanga yellow is also commonly used in slang to refer to gold. *Shar yum* thus means gold. In the lama's statement, shar lus emphasizes his conception of lus as Buddhist deity.
7. Uyanga's lamas also have a high regard for deer, which they describe as the most intelligent and beautiful animal of the forest. Carvings of deer adorn key Buddhist symbols in Mongolia. It is also common to see small figurines of deer on household altars in the city as well as in the countryside.
8. Buddhist specialists in the region generally use the expression *üiliin ürgüi yum* to denote the karmic condition of human action. In their usage, it does not necessarily refer to "good" or "bad" karmic actions. However, my various lay host families often used the expression to highlight behavior they considered questionable.
9. The daughter-in-law's mother lived nearby and was suffering financial hardship. However, as mentioned in chapter 1, the daughter-in-law had before asserted her autonomy from Ahaa's kin group and moved away with their only son. The daughter-in-law's failure to protect hishig was seen as yet another example of her unwillingness to be incorporated into Yagaanövgön's ail.
10. In Uyanga, many of the readings are in Mongolian, but there are also some in Tibetan. However, hardly any of the lamas can translate Tibetan. As discussed by Cook (2010, 113) in the context of Thai monastic meditation practices, "the experiences engendered by the use of Pali in chanting and meditation are not dependent upon understanding the meaning of the words." When they use a sacred, noncomprehensible language, the emotional experiences and processes of rationalization can jointly, without one's taking primacy over another, confirm the truth of their experiences and insights (see also Humphrey and Ujeed 2013, 265–277, and Hyer and Jagchid 1983, 8–9).
11. Bawden (1994) mentions *qara luus* (manuscript H 66) but does not elaborate further on its characteristics.
12. According to Grégory Delaplace (personal communication), the word *gai* is commonly used in funerary contexts when the possible causes behind a person's death are considered. He has emphasized that *gai* is used only to explain harms after they have occurred; it is not used in speculations about *potential* harms, as is the case in Uyanga. Bawden (1994, 72) also makes reference to "pathogenic agency" in his study of Mongolian conceptions of illness and death but unfortunately does not provide a Mongolian correlate for the English term.

Chapter 4. Polluted Money

1. For a contrastive example in which money earned from mining is regarded as fertile and benign, see Harris's 1989 discussion of the ways in which the Laymi of the Bolivian Andes conceptualize mineral wealth. In contrast to the ethnographic examples mentioned in the text, it is worth noting that in the Potosí region mining has been carried out for centuries and historically formed an integral part of both the state apparatus and the cosmic order.

2. For a related example of how jealousy and envy give rise to fears of hel am, yet in the highly different context of urban pawnbroking in Mongolia, see Højer 2012.
3. Child-bearing women are vulnerable not only to hel am but also to pollution. Empson (2011, 246–49) offers a detailed account of how hunting practices that entail wrongful acts and the bringing home of polluting carcasses can affect people, especially pregnant women.
4. After 1 January 1927, all state ministries and trade organizations in Mongolia were required to conduct transactions in the new state currency, and in order to limit the circulation of Chinese currency, the government decreed a restricted period of one-to-one exchange of Chinese currency.
5. The rate of inflation peaked in 1992 at 325 percent per year (Griffin 2003, 2).
6. Notable exceptions include the gold dinar coins that circulated during the Mongol Empire in the ilkhanate in Persia and the Chagatai khanate in modern-day Uzbekistan.
7. Buyanaa here refers to the mass of pure gold (its carat). Sometimes ninjas refer to 24k gold as 99, other times as 999. This variation in digits is common also in the European gold trade and belongs to the "millesimal fineness system," which denotes parts per thousand of pure metal in the alloy.
8. People usually store money at home in locked wooden storage boxes. Bank accounts are often described as similar to storage boxes but held by strangers. As such, they are seen to involve the containment of money and can potentially "swell up" with pollution if it holds gold money.
9. The noun *buzar* carries meanings such as "dirt," "filth," and "defilement" (Bawden 1997, 65). Considering the general cosmoeconomy invoked by buzar, I also translate the term as "pollution."
10. With regard to the association of blood with pollution, Swancutt (2012, 199) corroborates that there is a "widespread Inner Asian notion that menstrual blood, birth blood and blood from strange or unknown sources are polluting (Diemberger 1993:113–122, Humphrey 1996:181, see also 301–302)."
11. For an example of how Mongolian concerns with pollution produce a sense of mistrust, suspicion, and paranoia toward the Chinese more generally, see Frank Billé 2015.
12. Pollution can also be transferred onto other material objects, not just money. For example, in her account of Buryats in northeastern Mongolia Buyandelger (2013, 160) mentions that a boy fell ill "because of pollution from a silver bridle he had picked up in the wilderness a few days ago. The bridle needed to be returned to the place where it had been found." Material objects that have been dropped and lie on the ground are not considered alienable, passive, or neutral. As Empson (2007, 114, 135) has shown, there are intimate continuous relations between material objects and persons/intentions in Mongolia, rendering objects extensions of the intentionality of people (see also Humphrey 2002). As a result, any dropped object is likely to retain the intentions of its previous owner(s) and may be an object of misfortune. Moreover, objects are also vulnerable to and can be affected by the forces of the environment they are in. In this case, an object with no apparent black footprints can become troublesome because it has touched something polluted (see, for example, Højer's 2012 account of items held by pawnbrokers in Ulaanbaatar). Given the dynamic conceptualization of objects, lost objects are thus best left alone lest various sources of misfortune follow.
13. The term *altny chanj* consists of the substantivized Mongolian version of the English verb "to change." According to Højer (2012, 37), "this term only entered the Mongolian vocabulary in the 1990s with the introduction of the new market economy. At first it was used to refer to people involved in currency exchange, but nowadays the label can, at least in principle, be used for any kind of person involved in exchange-for-profit. In this general sense, a 'changer' is simply an intermediary who buys up goods to sell them on" (see also Yule 2007).

Chapter 5. Wealth and Devotion

1. The procession of the Maidar Buddha is an elaborate annual event and is carried out in many Mongolian and Buryat monasteries (for other descriptions see Humphrey and Ujeed 2013, 9–25; Jagchid and Hyer 1979: 126–27; Borjigin 2006, 45–47).
2. Given the different circumstances of my fieldwork with Uyanga's lamas, in which questions of illegality were not necessarily an issue, I was able to record most of my conversations with my hosts. Hence the longer statements that appear in this chapter, based on successive periods of fieldwork carried out between 2005 and 2011.
3. Central to various Buddhist traditions is the understanding of the "three ages of the Dharma," in which the third and last age is characterized by moral degeneration, strife, and natural disasters. In addition to the Mahayana Buddhist texts such as the *Diamond Sutra* and the *Lotus Sutra,* see also the *Lion's Roar on the*

Turning of the Wheel (Sanskrit *Cakkavatti-Sihanada Sutta*) in the *Dīgha Nikāya* collections, which describes humanity's decline with a prophecy of its eventual apocalyptic return.

4. The unique position of human beings for following the path to salvation, as described by the Dalai Lama, is debated within the Buddha's own teachings. In the *Lotus Sutra*, for example, it is made clear that the attainment of Buddhahood is available to all beings who follow the teachings, even to the epitome of evil, Devadatta, and the nagas protecting the Buddhist doctrine (Watson 1993, chap. 12).

5. Although there have been concerted, and often internationally backed, efforts to reinvigorate Mongolian Buddhism over the last few decades, the notion of "revitalization" might not describe adequately the heterogeneous processes that have taken place (see Abrahms-Kavunenko 2011).

6. In his analysis of the Tibetan soil ritual *Sa Chog*, Gardner (2006) argues that the ritual is not intended to avoid inflicting harm on the serpent but rather, alongside its veneration of nonhuman beings, to "maximise the violence." In Uyanga, the rituals not only address spirits that are different from the Tibetan soil serpent and not nearly as elaborate as those described by Gardner but also do not seem to bear any parallels to the Tibetan "symbolic stabbing wrapped in a wrathful subjugation narrative" (ibid., 11). Beyond the lamas' own commentary, the ritual orchestration itself that calls forth spirits and offers items most desirable to them, the patient waiting and the still unknown aftermath, ascertains the expected display of respect and veneration of other beings. It is thus more in line with Charles Ramble's (1996, 145) discussion of a harvest ritual in which sacred texts are recited in order to protect the cultivated fields from external harm.

7. Although the monks lend support to Humphrey's view of the successful espousal of socialist ideology in Mongolia, there are also numerous counterexamples. Apart from the fact that many people continued their religious practices, albeit in secrecy, during the socialist period, staunch defenders of socialism also incorporated traditional rituals into the socialist work regime. The former governor during the late socialist years in Uyanga, for example, recounts that they made offerings before laying the foundations of new buildings in the 1980s: "We had to ask the land for permission."

8. The ideological association between socialism and neoliberalism has also been discussed by Boyer and Yurchak (2010) in the context of a particular parody genre (*stiob*) that flourished in late socialist USSR and today's US late liberalism.

9. In referring to energy, Budlam is likely drawing on both popular Russian new age spirituality and Tantric Buddhist understandings of energy flowing through our bodies and the universe, creating and upholding our being. Through rituals and meditation we can contact and channel this energy, thereby achieving mindfulness and liberation from suffering.

10. In a similar vein, the Buddhist scholar David McMahan (2008, 149) writes, "That we live in a radically interconnected world has become a truism. Indeed, this age of internationalism and the internet might well be called the age of *inter*: there is nothing that is not interconnected, interdependent, interwoven, interlaced, interactive, or interfacing with something else to make it what it is. Thus, any religious tradition that can claim 'interdependence' as a central doctrine lays claim to timely cultural resonance and considerable cultural cachet." In his attempts to position Buddhism as a preeminently "modern religion," McMahan evinces a certain theory of globalization that is fraught with presumptions of unprecedented connectedness.

11. For other examples of how money and alms mediate social relationships within monastic communities, see Laidlaw 1995 and Cook 2008.

12. Despite his rather controversial legacy as a brutal conqueror, Chinggis Khaan (1162?–1227) is today a national hero in Mongolia, admired for having united the warring Mongol groups and founded the Mongol Empire. General Sühbaatar (1893–1923) is another national hero in contemporary Mongolia. He was the founder and first commander of the Mongolian armed forces in the 1921 revolution, supported by the Soviet Red Army in the fight against the theocratic ruler Bogd Khan and his allied Chinese forces.

Chapter 6. Trading Gold

1. Billiards (*bil'yard*) is very popular throughout Mongolia. Beyond the pool halls in Ulaanbaatar, there are often pool tables in market areas and in billiard gers in the provincial and district centers. In the mines of Uyanga, the altny chanj keeps the sticks and balls inside his ger, handing them over upon the payment of a small fee.

2. Since Bayasgalan and Battsetseg do not keep a formal record of their day-to-day financial transactions, the calculations are based partly on their own estimations, my observations, and gathered data.

3. The term *ersdeltei* occurs in the composite noun *ersdeltei naimaa,* which simply means a "venture." However, when it is used as an adjective, it can carry meanings such as "risky," "unsafe," and "precarious"—for example, in the expression *ersdeltei höröngö oruulalt* (an insecure investment)—yet can also be used to denote something as adventurous. In delineating the desired meaning of ersdeltei business, many people offered the synonym *ayuultai business,* meaning "dangerous business."
4. In line with Elizabeth Ferry's (2016, 83) insights on the intricacies of gold pricing, I am here referring to the London Gold Fix as an "[entity] whose main purpose is to express value (even if they sometimes do so badly)." I did question people about the London-offered gold price, but I did not encounter any curiosity or interest among my interlocutors in this topic. Despite its crucial implication in and for the Mongolian gold rush, the London price appeared as a kind of unchallengeable, already composed entity that was announced on TV and printed in newspapers (see also Muniesa 2007).
5. Given the intense and dynamic relations between humans and material objects, Empson (2011, 147–81) examines how Mongolian Buryats confront the challenges of maintaining boundaries and encouraging connections with certain objects, such as umbilical cords, pieces of hair, dresses, and cots. These artifacts are not static or inert but rather centrally implicated in the generation and harnessing of fortune.
6. For a detailed discussion of how the general hierarchization of social relations coexisted with a Soviet socialist ideology of equality among Buryats in two collective farms in Siberia during and after the USSR, see Humphrey 1983, 1998.
7. There are many parallels between Uyanga's big bosses and the local-level leaders in the neighboring Arhangai province during the socialist period as described by Daniel Rosenberg (1981). Rather than drawing first and foremost on their official roles, these leaders seem to excel at mobilizing their individual histories and acquired personal qualities to cement their positions of authority locally. Becoming recognized as a leader thus seems related to your ability to move beyond the official position and become a person in the sense of an "exemplar" (Humphrey 1995).
8. According to the first nationwide survey of ninja mining in Mongolia, the offered gold price in Uyanga in 2003 was only 9,200 MNT (Grayson et al. 2004, 57). In line with the London Gold Fixing, the local rise in gold price over a relatively short period has thus been immense and unmatched in other economic sectors of the country.
9. At this time, the national average monthly income per person was estimated to be around 34 USD (National Statistical Office of Mongolia 2006, 261). Although the profit margin for the petty gold traders was small, they were earning about three times the national average.
10. Empson (2011, 268–315) describes how arson attacks occurred repeatedly among the Buriads in Ashinga, Mongolia, during a time of increasing inequalities in wealth. The targets were often young, successful entrepreneurs who were running new businesses. Empson argues that rather than being a means by which people come to terms with modernity and increasing polarities in wealth, arson makes visible "more lingering senses of animosity" (ibid., 305) over the accumulation of wealth in private property as something owned by individuals (see also Humphrey 2002). The dangers associated with accumulating individually held wealth are thus not limited to the gold mining areas and their exceptional potential for profit making.
11. There is an important continuity in the elite positions held by Uyanga's big bosses during the socialist period and those held today. This is not a case in which people who were marginalized and disenfranchised in the past draw on new economic opportunities to reposition themselves in society. Rather, in line with Rosenberg (1981) and Humphrey (1983), the big bosses have held various leadership positions in the past and have been able to continue their positionality through the gold rush. Yet one should not assume a fully antagonistic relationship between leader and subordinate. As noted by Humphrey (1994, 23), during the socialist period there was a "structural encapsulation" whereby Mongolia and its various leaders were "held in thrall to a Soviet Union that operated from above and encompassed Mongolia's own internal hierarchy. . . . All Mongolians in 'power' also experienced the insult of subordination in their own spheres." While Russia today does not hold the same power over Mongolia, other forces, such as finance capital, continue processes of structural encapsulation.

GLOSSARY

ail household cluster
aimag regional province (highest administrative level below state)
altny chadvar a sort of "invisible dust" that rubs off gold during the process of extraction and can lead to a greed for gold (lit. power of gold)
altny chanj gold trader (lit. someone who changes gold)
altny gai a substance of affliction that can be redirected onto others through ritual practices (lit. the misfortune of gold)
altny möngö money earned from gold mining (lit. gold money)
ashig profit
bag county (smallest administrative level, below *sum*)
bayalag wealth
buzar cosmological pollution (lit. dirt, filth, impurity)
dahin törölt rebirth
deel Mongolian robe
enerji an affective quality that highlights money's capacity to become productive and profit-earning capital (lit. energy)
ersdeltei business risky business
ger yurt (nomadic tent)
hel am malicious gossip
hishig fortune, blessings
lam Buddhist lama (monk)
lus savdag water and earth spirits
ninja informal-sector gold miner
nutag homeland
Ongiin Golynhan Hödölgöön The Ongi River Movement, an environmental advocacy group founded to protect the Ongi River from destructive mining activities (lit. the Movement of the People of the Ongi River)
ovoo stone cairn (place of ritual practices)
shar yum gold (lit. yellow stuff)
shoroony gazar land of dust (refers to the mining areas generally)
shoroo ugaah to pan for gold (lit.to wash dirt)
sum district (intermediate administrative level above *bag* and below *aimag*)

süns soul

tom darga an illegal gold trader positioned above all the local traders who sells the gold in the capital and beyond (lit. big boss)

tögrög (MNT) Mongolian national currency (in 2005–6, 1 USD equaled approx. 1200 MNT)

zud disaster affecting livestock caused by several weather conditions

REFERENCES

Abrahms-Kavunenko, Saskia. 2011. "Improvising Tradition: Lay Buddhist Experiences in Cosmopolitan Ulaanbaatar." PhD diss., University of Western Australia.

Abramson, Allen, and Martin Holbraad. 2012. "Contemporary Cosmologies, Critical Reimaginings." *Religion and Society: Advances in Research* 3:35–50.

Adepoju, Aderanti. 2003. "Migration in West Africa." *Development* 46 (3): 37–41.

Akin, David, and Joel Robbins, eds. 1999. *Money and Modernity: State and Local Currencies in Melanesia.* Pittsburgh: University of Pittsburgh Press.

Aldama, Zigor. 2016. "Winners and Losers in Mongolia's Mining Gold Rush." *South China Morning Post (SCMP)*, April 29.

Andersson, Ake E., David F. Batten, Kiyoshi Kobayashi, and Kazuhiro Yoshikawa. 1993. *The Cosmo-Creative Society: Logistical Networks in a Dynamic Economy.* Düsseldorf: Springer-Verlag.

Appadurai, Arjun. 1986. Introduction: Commodities and the Politics of Value." In *The Social Life of Things: Commodities in a Cultural Perspective,* edited by Arjun Appadurai, 3–63. Cambridge: Cambridge University Press.

Ardyn Erh [People's Rights]. 2006. "Baigal' orchny said I. Erdenebaataryn baigal' 'hamgaalsan' tüüh" [A history of how Minister of Environment I. Erdenbaatar "protected" the environment]. March 3.

Atwood, Christopher P. 2004. *Encyclopedia of Mongolia and the Mongol Empire.* Bloomington: Indiana University Press.

Auty, Richard M., ed. 2001. *Resource Abundance and Economic Development.* Oxford: Oxford University Press.

Awehali, Brian. 2011. "Mongolia's Wilderness Threatened by Mining Boom." *Guardian,* January 11.

Ballard, Chris, and Glenn Banks. 2003. "Resource Wars: The Anthropology of Mining." *Annual Review of Anthropology* 32:287–313.

Bareja-Starzynska, Agata, and Hanna Havnevik. 2006. "A Preliminary Study of Buddhism in Present-Day Mongolia." In *Mongols from Country to City: Floating Boundaries, Pastoralism and City Life in the Mongol Lands,* edited by Ole Bruun and Li Narangoa, 212–36. Copenhagen: NIAS Press.

Barta, Patrick, and Jargal Byambasuren. 2007. "Big Dig: Mongolia Is Roiled by Miner's Huge Plans." *Wall Street Journal,* January 4.

Barth, Frederik. 1975. *Ritual and Knowledge among the Baktaman of New Guinea.* New Haven: Yale University Press.

Barthelemy, Pierre. 2013. "Mongol(d)ia—A Cursed Gold Rush?" *Political Bouillon,* January 10.

Basso, Keith H. 1996. *Wisdom Sits in Places: Landscape and Language among the Western Apache.* Albuquerque: University of New Mexico Press.

Batima, Punsalmaa, Luvsan Natsagdorj, and Nyamsurengyn Batnasan. 2008. "Vulnerability of Mongolia's Pastoralists to Climate Extremes and Changes." In *Climate Change and Vulnerability.* Vol. 2, edited by Neil Leary, Cecilia Conde, Jyoti Kulkarni, Anthony Nyong and Juan Pulhin, 67–87. London: Earthscan.

Bawden, Charles. 1994. "The Supernatural Elements in Sickness and Death according to Mongol Tradition, Part I." In *Confronting the Supernatural: Mongolian Traditional Ways and Means,* edited by Charles Bawden, 41–84. Wiesbaden: Harrassowitz Verlag.

——. 1997. *Mongolian-English Dictionary.* London: Kegan Paul International.

Bazuin, Gerritt. 2003. "Poverty-Driven Gold Rush Still Gathering Speed." *UB Post,* April 3.

Beidelman, Thomas O. 1989. "Agonistic Exchange: Homeric Reciprocity and the Heritage of Simmel and Mauss." *Cultural Anthropology* 4 (3): 227–59.

Benwell, Ann. 2006. "Facing Gender Challenges in Post-Socialist Mongolia." In *Mongols: From Country to City,* edited by Ole Bruun and Li Narangoa, 110–39. Copenhagen: NIAS Press.

Berdahl, Daphne. 1999. "'(N)Ostalgie' for the Present: Memory, Longing, and East German Things." *Ethnos* 64 (2): 192–211.

——. 2000. "Introduction: An Anthropology of Postsocialism." In *Altering States: Ethnographies of Transition in Eastern Europe and the Former Soviet Union,* edited by Daphne Berdahl, Matti Bunzl, and Martha Lampland, 1–13. Ann Arbor: University of Michigan Press.

Bernstein, Anya. 2011. "The Post-Soviet Treasure Hunt: Time, Space, and Necropolitics in Siberian Buddhism." *Comparative Studies in Society and History* 53:623–53.

Billé, Franck. 2013. "Indirect Interpellations: Hate Speech and 'Bad Subjects' in Mongolia." *Asian Anthropology* 12 (1): 3–19.

——. 2014. *Sinophobia: Anxiety, Violence, and the Making of Mongolian Identity.* Honolulu: University of Hawai'i Press.

Bloch, Maurice, and Jonathan Parry. 1989. "Introduction: Money and the Morality of Exchange." In *Money and the Morality of Exchange,* edited by Jonathan Parry and Maurice Bloch, 1–32. Cambridge: Cambridge University Press.

Boddy, Janice. 1989. *Wombs and Alien Spirits: Women, Men, and the Zār Cult in Northern Sudan.* Madison: University of Wisconsin Press.

Bohannan, Paul. 1955. "Some Principles of Exchange and Investment among the Tiv." *American Anthropologist* 57 (1): 60–70.

——. 1959. "The Impact of Money on an African Subsistence Economy." *Journal of Economic History* 19:491–503.

Borjigin, Uranchimeg. 2006. "Circulating Prophetic Texts." In *Time, Causality and Prophecy in the Mongolian Cultural Region,* edited by Rebecca Empson, 21–60. Folkestone, UK: Global Oriental.

Boyer, Pascal, and Alexei Yurchak. 2010. "American Stiob: Or, What Late-Socialist Aesthetics of Parody Reveal about Contemporary Political Culture in the West." *Current Anthropology* 25:179–221.

Branigan, Tania, and Dan Chung. 2014. "Mining in Mongolia." *Guardian,* April 23.

Bubandt, Nils, and Ton Otto. 2011. "Anthropology and the Predicaments of Holism." In *Experiments in Holism: Theory and Practice in Contemporary Anthropology,* edited by Ton Otto and Nils Bubandt, 1–15. Oxford: Wiley-Blackwell.

Bulag, Uradyn E. 1998. *Nationalism and Hybridity in Mongolia.* Oxford: Clarendon Press.

Buyandelger, Manduhai. 2007. "Dealing with Uncertainty: Shamans, Marginal Capitalism, and the Remaking of History in Postsocialist Mongolia." *American Ethnologist* 34:127–47.

———. 2013. *Tragic Spirits: Shamanism, Memory, and Gender in Contemporary Mongolia.* Chicago: University of Chicago Press.

Byambajav, Dalaibuyan. 2015. "The River Movements' Struggle in Mongolia." *Social Movement Studies* 14 (1): 92–97.

Campbell, Charles W. 1903. "Journeys in Mongolia." *Geographical Journal* 22 (5): 485–518.

Castells, Manuel. 2010. *The Power of Identity.* Vol. 2 of *The Information Age: Economy, Society and Culture.* Oxford: Wiley-Blackwell.

Castree, Noel. 2014. *Making Sense of Nature.* Abingdon, UK: Routledge.

Chabros, Krystyna. 1992. *Beckoning Fortune: A Study of the Mongol Dalalga Ritual.* Wiesbaden: Otto Harrassowitz.

Chimedsengee, Urantsatsral, Amber Cripps, Victoria Finlay, Guido Verboom, Munkhbaatar Batchuluun, and Byambajav Khunkhur. 2009. *Mongolyn burhany shashintnuudyn baigal hamgaalah üil ajillagaa* [Mongolian Buddhists protecting nature]. Ulaanbaatar, Mong.: Alliance of Religions and Conservation.

Clark, Jeremy. 1993. "Gold, Sex, and Pollution: Male Illness and Myth at Mt. Kare, Papua New Guinea." *American Ethnologist* 20 (4): 742–57.

Comaroff, Jean, and John L. Comaroff. 1999. "Occult Economies and the Violence of Abstraction: Notes from the South African Postcolony." *American Ethnologist* 26 (2): 279–303.

———. 2000. "Millennial Capitalism: First Thoughts on a Second Coming." *Public Culture* 12 (2): 291–343.

Cook, Joanna. 2008. "Alms, Money and Reciprocity: Buddhist Nuns as Mediators of Generalised Exchange in Thailand." *Anthropology in Action* 15 (3): 8–21.

———. 2010. *Meditation in Modern Buddhism: Renunciation and Change in Thai Monastic Life.* Cambridge: Cambridge University Press.

Cooper, Louise. 1995. *Wealth and Poverty in the Mongolian Pastoral Economy.* Research Report No. 11. Ulaanbaatar, Mong.: Institute of Development Studies at the University of Sussex, Institute of Animal Husbandry at the Mongolian Agricultural University (Mongolia), and Institute of Agricultural Economics (Mongolia).

Cribb, Joe. 2005. "Money as Metaphor: Money Is Justice, the Origins of Money and Coinage." *Numismatic Chronicle* 165:417–38.

Crump, Thomas. 2011. *The Phenomenon of Money.* London: Routledge & Kegan Paul.

Crutzen, Paul. 2002. "Geology of Mankind." *Nature* 415 (6867): 23–23.

da Col, Giovanni. 2012. "The Poisoner and the Parasite: Cosmoeconomics, Fear, and Hospitality among Dechen Tibetans." *Journal of the Royal Anthropological Institute* 18 (S1): S175–S195.

Dalaibuyan, Byambajav. 2012. "Formal and Informal Networks in Post-Socialist Mongolia: Access, Uses, and Inequalities." In *Change in Democratic Mongolia: Social Relations, Health, Mobile Pastoralism, and Mining,* edited by Julian Dierkes, 31–54. Leiden, Neth.: Brill.

Dalai Lama. 1994. *The Way to Freedom: Core Teachings of Tibetan Buddhism.* San Francisco: Harper Collins.

Day, John. 1999. *Money and Finance in the Age of Merchant Capitalism 1200–1800.* Oxford: Blackwell.

De Boeck, Filip. 1999. "Domesticating Diamonds and Dollars: Identity, Expenditure and Sharing in Southwestern Zaire (1984–1997)." In *Globalization and Identity: Dialectics of Flow and Closure,* edited by Birgit Meyer and Peter Geschiere, 177–210. Oxford: Blackwell.

De la Cadena, Marisol. 2010. "Indigenous Cosmopolitics in the Andes: Conceptual Reflections beyond 'Politics.'" *Cultural Anthropology* 25 (2): 334–70.

Delaplace, Gregory. 2009. "A Sheep Herder's Rage: Silence and Grief in Contemporary Mongolia." *Ethnos* 74 (4): 514–34.

———. 2010. "Chinese Ghosts in Mongolia." *Inner Asia* 12:127–41.

Descola, Philippe. 1996. "Constructing Natures." In *Nature and Society: Anthropological Perspectives,* edited by Philippe Descola and Gísli Pálsson, 82–102. London: Routledge.

Diemberger, Hildegaard. 1993. "Blood, Sperm, Soul and the Mountain: Gender Relations, Kinship and Cosmovision among the Khumbo (NE Nepal)." In *Gendered Anthropology,* edited by Teresa del Valle, 88–127. London: Routledge.

Donham, Donald L. 2011. *Violence in a Time of Liberation: Murder and Ethnicity at a South African Gold Mine, 1994.* Durham, NC: Duke University Press.

Dudley,Nigel, Liza Higgins-Zogib, and Stephanie Mansourian. 2005. *Beyond Belief: Linking Faith and Protected Areas to Support Biodiversity Conservation.* Gland, Switz.: World Wildlife Fund and the Alliance of Religions and Conservation.

Durkheim, Émile. 1893. *De la Division du Travail Social: Étude sur l'organisation des sociétés supérieures* [The division of labor in society: Research on the organization of superior societies]. Paris: Presses Universitaires de France.

Economist. 2012. "Booming Mongolia: Mine, All Mine." January 21, 23–25.

Eifler, Mark A. 2005. *Gold Rush Capitalists: Greed and Growth in Sacramento.* Albuquerque: University of New Mexico Press.

Eiss, Paul K., and David Pedersen. 2002. "Introduction: Values of Value." *Cultural Anthropology* 17 (3): 283–90.

Eliade, Mircea. (1951) 2004. *Shamanism: Archaic Techniques of Ecstasy.* Reprint, Princeton: Princeton University Press. Page references to 2004 edition.

Elson, Diane. 1979. *Value: The Representation of Labour in Capitalism.* London: CSE Books.

Elverskog, Johan. 2006. "Two Buddhisms in Contemporary Mongolia." *Contemporary Buddhism* 7:29–46.

Empson, Rebecca. 2003. "Integrating Transformations: A Study of Children and Daughters-in-Law in a New Approach to Mongolian Kinship." PhD diss., Cambridge University.

———. 2007. "Separating and Containing People and Things in Mongolia." In *Thinking through Things: Theorising Artefacts Ethnographically,* edited by Amiria Henare, Martin Holbraad, and Sari Wastell, 113–40. London: Routledge.

———. 2011. *Harnessing Fortune: Personhood, Memory and Place in Mongolia.* Oxford: Oxford University Press and British Academy.

———. 2012. "The Dangers of Excess: Accumulating and Dispersing Fortune in Mongolia." *Social Analysis* 56 (1): 117–32.

Enhbat, Chuluundorjiin. 2010. *Foreign Coins, Banknotes and Other Media of Exchange Used in Mongolia.* Ulaanbaatar, Mong.: Munhiin Useg Group.

Erel Kompani. 1994. *Uyanga orchmyn geologiintTogtots ashigt maltmal* [The geological mineral structure of the Uyanga area]. Mineral Report No. 4756. Ulaanbaatar, Mong.: Ministry of Thermal Energy and Geological Mining.

Evans-Pritchard, Edward Evan. 1937. *Witchcraft, Oracles and Magic among the Azande.* Oxford: Clarendon Press.

Feldstein, Martin. 2009. "Is Gold a Good Hedge?" Project Syndicate, December 13.

Ferguson, James. 1992. "The Cultural Topography of Wealth: Commodity Paths and the Structure of Property in Rural Lesotho." *American Anthropologist* 94 (1): 55–73.

Ferry, Elizabeth E. 2002. "Inalienable Commodities: The Production and Circulation of Silver and Patrimony in a Mexican Mining Cooperative." *Cultural Anthropology* 17 (3): 331–58.

——. 2005. *Not Ours Alone: Patrimony, Value, and Collectivity in Contemporary Mexico*. New York: Columbia University Press.

——. 2016. "Gold Prices as Material-Social Actors: The Case of the London Gold Fix." *Extractive Industries and Society* 3 (1): 82–85.

Fiéloux, Michèle. 1980. *Les Sentiers de la Nuit: Les Migrations Rurales Lobi de la Haute-Volta vers la Côte d'Ivoire* [The Paths of the Nights: The rural migrations of the Lobi from the Upper Volta toward the Ivory Coast]. Travaux et Documents 110. Paris: Office de la Recherche Scientifique et Technique d'Outre Mer (ORSTOM).

Firth, Raymond. 1929. *Primitive Economics of the New Zealand Maori*. London: Routledge.

Foster, Robert J. 1990. "Value without Equivalence: Exchange and Replacement in a Melanesian Society." *Man*, n.s., 25 (1): 54–69.

——. 1998. "Your Money, Our Money, the Government's Money: Finance and Fetishism in Melanesia." In *Border Fetishisms: Material Objects in Unstable Places*, edited by Patricia Spyer, 60–90. London: Routledge.

Foucault, Michel. 1970. *The Order of Things: An Archaeology of the Human Sciences*. New York: Vintage.

Fourcade, Marion. 2011. "Cents and Sensibility: Economic Valuation and the Nature of 'Nature.'" *American Journal of Sociology* 116 (6): 1721–77.

Friedman, Milton. 1962. *Capitalism and Freedom*. Chicago: Chicago University Press.

Friedman, Milton, and Anna Jacobson Schwartz. 1963. *Monetary History of the United States: 1867–1960*. Princeton: Princeton University Press.

Frost, Lionel. 2010. "'Metallic Nerves': San Francisco and Its Hinterland during and after the Gold Rush." *Australian Economic History Review* 50 (2): 129–47.

Gampopa, Je. 1995. *Gems of Dharma, Jewels of Freedom*. Translated by Ken Holmes and Katia Holmes. Forres, Scot.: Altea Publishing.

Gantulga, Mönh-Erdene. 2011. *Burzaij baina uu? Ninja nar, tednii zohion baiguulalt hiigeed am zuulga* [Does it branch out? Ninjas and their social organization]. Ulaanbaatar, Mong.: Meeting Point Press.

Gardner, A. 2006. "The *Sa Chog*: Violence and Veneration in a Tibetan Soil Ritual." *Études mongoles et sibériennes, centrasiatiques et tibétaines* 36–37:283–323.

Gell, Alfred. 1992. *The Anthropology of Time: Cultural Constructions of Temporal Maps and Images*. Oxford: Berg.

Geshiere, Peter. 1997. *Modernity of Witchcraft: Politics and the Occult in Post-Colonial Africa*. Charlottesville: University Press of Virginia.

GFMS. 2004. "New GFMS Report Reviews the Key Drivers behind the Recent Exploration Boom in Mongolia." Press release, August 19.

Gilder, George F. 2013. *Knowledge and Power: The Information Theory of Capitalism and How It Is Revolutionizing Our World*. Washington, DC: Regnery.

Gilmour, James. 1883. *Among the Mongols*. London: Religious Tract Society.

Godoy, Ricardo. 1985. "Mining: Anthropological Perspectives." *Annual Review of Anthropology* 14:199–217.

Golub, Alex. 2014. *Leviathans at the Gold Mine: Creating Indigenous and Corporate Actors in Papua New Guinea*. Durham, NC: Duke University Press.

Graeber, David. 1996. "Beads and Money: Notes towards a Theory of Wealth and Power." *American Ethnologist* 23 (1): 4–24.

——. 2013. "It Is Value That Brings Universes into Being." *HAU: Journal of Ethnographic Theory* 3 (2): 219–43.

Grainger, David. 2003. "The Great Mongolian Gold Rush: The Land of Genghis Khan Has the Biggest Mining Find in a Very Long Time." *Fortune*, December 22.

Grayson, Robin. 2006. *The Gold Miners Book—Manual for Miners, Investors, Regulators and Environmentalists: Best Available Techniques for Placer Gold Miners*. Ulaanbaatar, Mong.: Eco-Minex International.

——. 2007. "Anatomy of the People's Gold Rush in Modern Mongolia." *World Placer Journal* 7:1–66.

Grayson, Robin, and Baatar Tumenbayar. 2005. "The Role of Placer Mining Companies in the State-Sponsored Gold Rush in Mongolia." *World Placer Journal* 5:1–36.

Grayson, Robin, Tsevel Delgertsoo, William Murray, Baatar Tümenbayar, Minjin Batbayar, Urtnasan Tuul, Dashzeveg Bayarbat, and Chimed Erdene-Baatar. 2004. "The People's Gold Rush in Mongolia: The Rise of the 'Ninja' Phenomenon." *World Placer Journal* 4:1–112.

Gregory, Chris A. 1997. *Savage Money: The Anthropology and Politics of Commodity Exchange*. Amsterdam: Harwood Academic Publishers.

Griffin, Keith, ed. 2003. *Poverty Reduction in Mongolia*. Canberra, Aus.: Asia Pacific Press.

Gudeman, Stephen. 2001. *The Anthropology of Economy: Community, Market, and Culture*. Malden, UK: Blackwell.

Guyer, Jane I. 1993. "Wealth in People and Self-Realization in Equatorial Africa." *Man*, n.s., 28 (2): 243–65.

——. 1995. "Introduction: The Currency and Its Dynamics." In *Money Matters: Instability, Values and Social Payments in the Modern History of West African Communities*, edited by Jane I. Guyer, 1–33. London: James Currey.

——. 2004. *Marginal Gains: Monetary Transactions in Atlantic Africa*. Chicago: University of Chicago Press.

Hamayon, Roberte. 1987. "Abuse of the Father, Abuse of the Husband: A Comparative Analysis of Two Buryat Myths of Ethnic Origin." In *Synkretismus in den Religionen Zentralasiens*, edited by Walther Heissig and Hans J. Klimkeit, 91–107. Wiesbaden: Otto Harrossowitz.

Hamayon, Roberte, and Namtcha Bassanoff. 1973. "De la Difficulté d'Être une Belle-Fille" [The difficulty of being a daughter-in-law]. *Études Mongoles* 4:7–74.

Hann, Chris. 2010. "Moral Economy." In *The Human Economy*, edited by Keith Hart, Jean-Louis Laville, and Antonio David Cattani, 187–99. Cambridge: Polity Press.

Haraway, Donna J. 2008. *When Species Meet*. Minneapolis: University of Minnesota Press.

Hardenberg, Donata. 2008. "The Battle for Mongolia's Resources." Al Jazeera, July 3.

Hart, Keith. 1986. "Heads or Tails: Two Sides of the Coin." *Man*, n.s., 21 (4): 637–56.

——. 2001. *Money in an Unequal World*. London: Textere.

——. 2007. "Money Is Always Personal and Impersonal." *Anthropology Today* 23 (5): 12–16.

Haslund, Henning. 1949. *Mongolian Journey*. Translated by F. H. Lyon. London: Routledge & Kegan Paul.

Healey, Christopher J. 1985. "Pigs, Cassowaries, and the Gift of the Flesh: A Symbolic Triad in Maring Cosmology." *Ethnology* 24 (3): 153–65.

——. 1988. "Culture as Transformed Disorder: Cosmological Evocations among the Maring." *Oceania* 59 (2): 106–22.

Heissig, Walther. 1980. *The Religions of Mongolia*. Berkeley: University of California Press.

High, Mette M. 2008a. "Dangerous Fortunes: Wealth and Patriarchy in the Mongolian Informal Gold Mining Economy." PhD diss., University of Cambridge.

——. 2008b. "Wealth and Envy in the Mongolian Gold Mines." *Cambridge Anthropology* 27 (3): 1–18.

——. 2012. "The Cultural Logics of Illegality: Living Outside the Law in the Mongolian Gold Mines." In *Contemporary Mongolia: Transitions and Development*, edited by Julian Dierkes, 249–70. Leiden, Neth.: Brill Publishers.

High, Mette M. 2013. "Believing in Spirits, Doubting the Cosmos: Religious Reflexivity in the Mongolian Gold Mines." In *Ethnographies of Doubt: Faith and Uncertainty in Contemporary Societies*, edited by M. Pelkmans, 59–84. London: I. B.Tauris.

——. 2016a. "Human Predation and Animal Sociality: The Transformational Agency of 'Wolf People' in Mongolia." In *Animals Out of Place: Cryptozoology in Anthropological Perspective*, edited by Samantha Hurn, 107–19. London: Routledge.

——. 2016b. "A Question of Ethics: The Creative Orthodoxy of Buddhist Monks in the Mongolian Gold Rush." *Ethnos*. http://dx.doi.org/10.1080/00141844.2016.1140215.

High, Mette M., and Jonathan Schlesinger. 2010. "Rulers and Rascals: The Politics of Gold in Qing Mongolian History." *Central Asian Survey* 29:289–304.

Hogendorn, Jan, and Marion Johnson. 2003. *The Shell Money of the Slave Trade*. Cambridge: Cambridge University Press.

Højer, Lars. 2004. "The Anti-Social Contract: Enmity and Suspicion in Northern Mongolia." *Cambridge Anthropology* 24:41–63.

——. 2009. "Absent Powers: Magic and Loss in Post-Socialist Mongolia." *Journal of the Royal Anthropological Institute* 15 (3): 575–91.

——. 2012. "The Spirit of Business: Pawnshops in Ulaanbaatar." *Social Anthropology* 20 (1): 34–49.

Hook, Leslie. 2011a. "Boomtown Mongolia." *Financial Times*, July 8.

——. 2011b. "The Pink House of Equities." *Financial Times*, January 27.

——. 2012a. "Mongolia: Can't Live with China, Can't Live without China. *Financial Times*, September 13.

——. 2012b: "Mongolia's Boom Economy Risks Overheating." *Financial Times*, February 27.

Hugh-Jones, Stephen. 1979. *The Palm and the Pleiades: Initiation and Cosmology in Northwest Amazonia*. Cambridge Studies in Social and Cultural Anthropology. Cambridge: Cambridge University Press.

Humphrey, Caroline. 1974. "Inside a Mongol Tent." *New Society* 30 (630): 273–75.

——. 1978. "Women, Taboo and the Suppression of Attention." In *Defining Females: The Nature of Women in Society*, edited by S. Ardener, 89–108. London: Croom Helm.

——. 1979. "Do Women Labour in a Worker's State? Women in 'Karl Marx Kolkhoz,' Barguzin Region, Buryat ASSR." *Cambridge Anthropology* 5 (2):1–17.

——. 1983. *Karl Marx Collective: Economy, Society and Religion in a Siberian Collective Farm*. Cambridge: Cambridge University Press.

——. 1992. "The Moral Authority of the Past in Post-Socialist Mongolia." *Religion, State and Society* 20:375–389.

——. 1993. "Women and Ideology in Hierarchical Societies in East Asia." In *Persons and Powers of Women in Diverse Cultures*, edited by S. Ardener, 173–92. Oxford: Berg.

——. 1994. "Remembering an 'Enemy': The Bogd Khaan in Twentieth-Century Mongolia." In *Memory, History, and Opposition under State Socialism*, edited by Rubie S. Watson, 21–44. Santa Fe: School of American Research Press.

——. 1995. "Chiefly and Shamanist Landscapes in Mongolia." In *The Anthropology of Landscape: Perspectives on Place and Space*, edited by Eric Hirsch and Michael O'Hanlon, 135–62. Oxford: Clarendon Press.

——. 1998. *Marx Went Away but Karl Stayed Behind.* Ann Arbor: University of Michigan Press.

——. 2000. "Rethinking Bribery in Contemporary Russia." In *Bribery and Blat in Russia: Negotiating Reciprocity from the Middle Ages to the 1990s,* edited by Stephen Lovell, Alena Ledeneva, and Andrei Rogachevskii, 241–56. Basingstoke, UK: Macmillan.

——. 2002. "Rituals of Death as a Context for Understanding Personal Property in Socialist Mongolia." *Journal of the Royal Anthropological Institute,* n.s., 8:65–87.

——. 2012. "Favors and 'Normal Heroes': The Case of Postsocialist Higher Education." *HAU: Journal of Ethnographic Theory* 2 (2): 22–41.

Humphrey, Caroline, and Urgunge Onon. 1996. *Shamans and Elders: Experience, Knowledge, and Power among the Daur Mongols.* Oxford: Clarendon Press.

Humphrey, Caroline, and Hurelbaatar Ujeed. 2013. *A Monastery in Time: The Making of Mongolian Buddhism.* Chicago: University of Chicago Press.

Hyer, Paul, and Sechin Jagchid. 1983. *A Mongolian Living Buddha: Biography of the Kanjurwa Khutughtu.* Albany: State University of New York Press.

International Labour Organisation. 2004. *Baseline Survey on Informal Gold Mining in Bornuur and Zaamar Soums of Tuv Aimag.* Ulaanbaatar, Mong.: International Labour Organisation, Mongolian Employers' Federation, and Population Teaching and Research Centre.

Isakova, Asel, Alexander Plekhanov, and Jeromin Zettelmeyer. 2012. "Managing Mongolia's Resource Boom." Working Paper No. 138. European Bank for Reconstruction and Development, London.

James, William. (1909) 1996. *A Pluralistic Universe.* Lincoln: University of Nebraska Press.

Jagchid, Sechin, and Paul Hyer. 1979. *Mongolia's Culture and Society.* Boulder, CO: Westview Press.

Janzen, Jörg. 2005. *Artisanal Gold Mining in Bornuur Sum/Tuv Aimag and Sharyn Gol Sum/Darkhan Uul Aimag.* Ulaanbaatar, Mong.: Swiss Development and Cooperation Agency.

Jønsson, Jesper Bosse, and Deborah Fahy Bryceson. 2009. "Rushing for Gold: Mobility and Small-Scale Mining in East Africa." *Development and Change* 40 (2): 249–79.

Jung, Maureen A. 1999. "Capitalism Comes to the Diggings: From Gold Rush Adventure to Corporate Enterprise." In *A Golden State: Mining and Economic Development in Gold Rush California,* edited by James J. Rawls, Richard J. Orsi, and Marlene Smith-Baranzini, 52–77. Berkeley: University of California Press.

Kaplonski, Christopher. 2003. "Thirty Thousand Bullets: Remembering Political Repression in Mongolia." In *Historical Injustice and Democratic Transition in Eastern Asia and Northern Europe: Ghosts at the Table of Democracy,* edited by Kenneth Christie and Robert Cribb, 155–68. London: Routledge Curzon.

——. 2008. "Prelude to Violence: Show Trials and State Power in 1930s Mongolia." *American Ethnologist* 35(2): 321-37.

Keane, Webb. 2001. "Money Is No Object: Materiality, Desire, and Modernity in an Indonesian Society." In *The Empire of Things: Regimes of Value and Material Culture,* edited by Fred Myers, 65–90. Santa Fe: School of American Research Press.

——. 2008. "Market, Materiality and Moral Metalanguage." *Anthropological Theory* 8 (1): 27–42.

Keynes, John Maynard. 1982. "Ancient Currencies." In *The Collected Writing of John Maynard Keynes,* edited by Donald Moggridge. London: Macmillan.

Kirsch, Stuart. 2007. "Indigenous Movements and the Risks of Counterglobalization: Tracking the Campaign against Papua New Guinea's Ok Tedi Mine." *American Ethnologist* 34 (2): 303–21.

Kogan, L. A. 1992. "The Philosophy of N. F. Fedorov." *Russian Studies in Philosophy* 30 (4): 7–27.

Kohn, Eduardo. 2013. *How Forests Think: Toward an Anthropology beyond the Human*. Berkeley: University of California Press

Kopytoff, Igor. 1986. "The Cultural Biography of Things: Commoditization as Process." In *The Social Life of Things: Commodities in a Cultural Perspective*, edited by Arjun Appadurai, 64–94. Cambridge: Cambridge University Press.

Kristensen, Benedikte Møller. 2007. "The Human Perspective." *Inner Asia* 9 (2): 275–89.

Kohn, Michael. 2011: "Mongolian Herder on Mission to Tackle Mining Firms." *Agence France-Presse*, July 6.

Kuhn, Philip A. 1990. *Soulstealers: The Chinese Sorcery Scare of 1768*. Cambridge, MA: Harvard University Press.

Labonne, Beatrice. 1996. "Artisanal Mining: An Economic Stepping Stone for Women." *Natural Resources Forum* 20 (2): 117–22.

Lacaze, Gaëlle. 2006. "La Notion de Technique du Corps Appliquée à l'Étude des Mongols." *Le Portique: Philosophie et Sciences Humaines* 17:151–65.

Laidlaw, James. 1995. *Riches and Renunciation: Religion, Economy, and Society among the Jains*. Oxford: Clarendon Press.

——. 2014. *The Subject of Virtue: An Anthropology of Ethics and Freedom*. Cambridge: Cambridge University Press.

Langfitt, Frank. 2012. *Mongolia Booms*, four-part series on National Public Radio, May 21–24.

Larson, Frans August. 1930. *Larson—Duke of Mongolia*. Boston: Little, Brown.

Latour, Bruno. 1993. *We Have Never Been Modern*. Cambridge, MA: Harvard University Press.

——. 2004. "Whose Cosmos, Which Cosmopolitics? Comments on the Peace Terms of Ulrich Beck." *Common Knowledge* 10 (3): 450–62.

——. 2009. *Politics of Nature: How to Bring the Sciences into Democracy*. Cambridge, MA: Harvard University Press.

——. 2013. Gifford Lectures, University of Edinburgh. http://www.ed.ac.uk/arts-humanities-soc-sci/news-events/lectures/gifford-lectures/archive/series-2012-2013/bruno-latour.

——. 2014. "Anthropology at the Time of the Anthropocene—A Personal View of What Is to Be Studied." Distinguished lecture delivered at the American Anthropological Association annual meeting, Washington.

Latour, Bruno, and Vincent Antonin Lépinay. 2009. *The Science of Passionate Interests: An Introduction to Gabriel Tarde's Economic Anthropology*. Chicago: Prickly Paradigm Press.

Lemon, Alaina. 1998. "'Your Eyes Are Green Like Dollars': Counterfeit Cash, National Substance, and Currency Apartheid in 1990s Russia." *Cultural Anthropology* 13 (1): 22–55.

Lévi-Strauss, Claude. 1969. *The Elementary Structures of Kinship*. Boston: Beacon Press.

Lewis, Ioan Myrddin. 2003. *Ecstatic Religion: A Study of Shamanism and Spirit Possession*. London: Routledge.

Lindskog, Benedikte V. 2000. "We Are All Insects on the Back of Our Motherland." Master's thesis, University of Oslo.

Lkhamsuren, M. 1982, "Manpower Policy and Planning in the Mongolian People's Republic." *International Labour Review* 121: 469–80.

Lomnitz, Claudio. 2003. "Times of Crisis: Historicity, Sacrifice, and the Spectacle of Debacle in Mexico City." *Public Culture* 15 (1): 127–47.

Lorimer, Jamie. 2012. "Multinatural Geographies for the Anthropocene." *Progress in Human Geography* 36 (5): 593–612.

Lovgren, Stefan. "Mongolia Gold Rush Destroying Rivers, Nomadic Lives." *National Geographic News*, October 17.

Luehrmann, Sonja. 2011. *Secularism Soviet Style: Teaching Atheism and Religion in a Volga Republic*. Bloomington: Indiana University Press.

Luning, Sabine. 2009. "Gold in Burkina Faso: A Wealth of Poison and Promise." In *Traditions on the Move: Essays in Honor of Jarich Oosten*, edited by Jan Jansen, Sabine Luning, and Erik de Maaker, 117–37. Amsterdam: Rozenberg.

Mahmood, Saba. 2005. *Politics of Piety: The Islamic Revival and the Feminist Subject*. Princeton: Princeton University Press.

Malinowski, Bronislaw. 1922. *Argonauts of the Western Pacific: An Account of Native Enterprise and Adventure in the Archipelagoes of Melanesian New Guinea*. London: G. Routledge and Sons.

Marx, Karl. 1932. *Economic and Philosophic Manuscripts of 1844*. Moscow: Progress Publishers.

——. (1867) 1967. *Capital: A Critique of Political Economy*. Vol. 1, *A Critical Analysis of Capitalist Production*. Reprint, New York: International Publishers.

Maurer, Bill. 2006. "The Anthropology of Money." *Annual Review of Anthropology* 35:15–35.

——. 2008. "Resocializing Finance? Or Dressing It in Mufti? Calculating Alternatives for Cultural Economies." *Journal of Cultural Economy* 1 (1): 65–78.

——. 2009. "Moral Economies, Economic Moralities: Consider the Possibilities!" In *Economics and Morality: Anthropological Approaches*, edited by Katherine E. Browne and B. Lynne Milgram, 257–70. Lanham, MD: AltaMira Press.

Mauss, Marcel. 1925. *Essai sur le Don: Forme et Raison de l'Échange dans les Sociétés Archaïques*. Paris: L'Année Sociologique.

McMahan, David L. 2008. *The Making of Buddhist Modernism*. Oxford: Oxford University Press.

Mills, Edwin W. 1929. "Gold Mining in Outer Mongolia." *Mining Journal (London)* 165 (4891): 399–402.

Mimica, Jadran. 2010. "Un/Knowing and the Practice of Ethnography: A Reflection on Some Western Cosmo-Ontological Notions and Their Anthropological Application." *Anthropological Theory* 10 (3): 203–28.

MRAM (Mineral Resources Authority of Mongolia). 2012. *Ashigt maltmalyn gazryn 2008–2012 ony üil ajillagaany tuhai* [Concerning the operations of the Mineral Resources Authority in the years 2008–2012]. Ulaanbaatar, Mong.: Mineral Resources Authority of Mongolia.

Mintz, Sidney. 2007. "Currency Problems in Eighteenth Century Jamaica and Gresham's Law." In *Process and Patterns in Culture: Essays in Honour of Julian H. Steward*, edited by Robert A. Manners, 248–65. New Brunswick, NJ: Transaction Publishers.

Mongolian Business Development Agency (MBDA). 2003. *Ninja Gold Miners of Mongolia: Assistance to Policy Formulation for the Informal Gold Mining Sub-Sector in Mongolia*. Ulaanbaatar, Mong.: Mongolian Business Development Agency.

Mongol Ulsyn Ih Hural (State Great Hural [Parliament] of Mongolia). 1997. Mongol Ulsyn Huul' Ashigt Maltmalyn Tuhai [Minerals Law of Mongolia], adopted June 5.

——. 2006. Mongol Ulsyn Huul' Ashigt Maltmalyn Tuhai, Shinechilsen Nairuulga [Minerals Law of Mongolia, amended], adopted July 8.

——. 2008. Ashigt Maltmal Gar Üildverleliin Argaar Olborloh Tuhai [Artisanal and Small-Scale Mining Operations, Temporary Regulation], adopted April 1.

Montagu, Ivor. 1956. *Land of Blue Sky—A Portrait of Modern Mongolia*. London: Dennis Dobson.

Moore, Henrietta L. 2006. "The Future of Gender or the End of a Brilliant Career?" In *Feminist Anthropology: Past, Present and Future*, edited by Pamela L. Geller and Miranda K. Stockett, 23–42. Philadelphia: University of Pennsylvania Press.

Moore, Roland S. 1994. "Metaphors of Encroachment: Hunting for Wolves on a Central Greek Mountain." *Anthropological Quarterly* 67 (2): 81–88.

Muniesa, Fabian. 2007. "Market Technologies and the Pragmatics of Prices." *Economy and Society* 36 (3): 377–95.

Munn, Nancy D. 1986. *The Fame of Gawa: A Symbolic Study of Value Transformation in a Massim (Papua New Guinea) Society*. Cambridge: Cambridge University Press.

Murray, William, and Robin Grayson. 2003. *Overview of Artisanal Mining Activity in Mongolia.* Report prepared at the request of the World Bank. Ulaanbaatar.

Myers, Fred R. 2001. "Introduction: The Empire of Things." In *The Empire of Things: Regimes of Value and Material Culture*, edited by Fred R. Meyers, 3–61. Santa Fe: School of American Research Press.

Nadasdy, Paul. 2007. "The Gift in the Animal: The Ontology of Hunting and Human-Animal Sociality." *American Ethnologist* 34 (1): 25–43.

Namjil, Ganbat. 2008. "Mongolians Vote in Parliamentary Elections; Sharing Mineral Wealth a Major Issue." *International Herald Tribune*, June 29.

Nash, June. 1979. *We Eat the Mines and the Mines Eat Us: Dependency and Exploitation in Bolivian Tin Mines*. New York: Columbia University Press.

———. 2001. "Cultural Resistance and Class Consciousness in Bolivian Tin-Mining Communities." In *Power and Popular Protest: Latin American Social Movements*, edited by S. Eckstein, 182–202. Berkeley: University of California Press.

National Statistical Office of Mongolia. 2003. *Mongolian Statistical Yearbook 2002*. Ulaanbaatar, Mong.: National Statistics Office.

———. 2006. *Mongolian Statistical Yearbook 2005*. Ulaanbaatar, Mong.: National Statistics Office.

National Statistics Office and United Nations Development Programme (UNDP). 1999. *Living Standards Measurement Survey*. Ulaanbaatar, Mong.: National Statistics Office.

Nattier, Jan. 1991. *Once upon a Future Time: Studies in a Buddhist Prophecy of Decline*. Berkeley: Asian Humanities Press.

Nebesky-Wojkowitz, Réne de. 1956. *Oracles and Demons of Tibet: The Cult and Iconography of the Tibetan Protective Deities*. The Hague: Mouton.

Nordby, Judith. 1987. "The Mongolian People's Republic in the 1980s: Continuity and Change." *Journal of Communist Studies* 3:113–31.

Novaya Gazeta. 2011. "Wake Up, Mongolians!" April 29.

Ödriin Sonin [*Daily News*]. 2007. [Gold mining companies fail to engage in environmental rehabilitation]. August 30.

Ong, Aihwa. 1987. *Spirits of Resistance and Capitalist Discipline: Factory Women in Malaysia*. Albany: SUNY Press.

Ortiz, Horacio. 2013. "Financial Value: Economic, Moral, Political, Global." *HAU: Journal of Ethnographic Theory* 3 (1): 64–79.

Orwell, George. 1946. "In Front of Your Nose." *Tribune* (London), March 22.

Pedersen, Morten A. 2001. "Totemism, Animism and North Asian Indigenous Ontologies." *Journal of the Royal Anthropological Institute* 7:411–27.

———. 2011. *Not Quite Shamans: Spirit Worlds and Political Lives in Northern Mongolia*. Ithaca, NY: Cornell University Press.

Pedersen, Morten, Rebecca Empson, and Caroline Humphrey. 2007. "Editorial Introduction: Inner Asian Perspectivisms." *Inner Asia* 9 (2): 141–52.

Pedersen, Morten A., and Lars Højer. 2008. "Lost in Transition: Fuzzy Property and Leaky Selves in Ulaanbaatar." *Ethnos* 73 (1): 73–96.

Pilling, David. 2012. "Steady—Mongolia Is Not Yet the New Qatar." *Financial Times,* May 30.

Quijada, Justine Buck. 2012. "Soviet Science and Post-Soviet Faith: Etigelov's Imperishable Body." *American Ethnologist* 39:138–54.

Rajak, Dinah. 2011. *In Good Company: An Anatomy of Corporate Social Responsibility.* Stanford: Stanford University Press.

Ramble, Charles. 1996. "Patterns of Places." In *Reflections of the Mountain: Essays on the History and Social Meaning of the Mountain Cult in Tibet and the Himalaya,* edited by A. M. Blondeau and E. Steinkellner, 141–56. Vienna: Verlag der Österreichischen Akademie der Wissenschaften.

Rawls, James J. 1999. "A Golden State: An Introduction." In *A Golden State: Mining and Economic Development in Gold Rush California,* edited by James J. Rawls, Richard J. Orsi, and Marlene Smith-Baranzini, 1–23. Berkeley: University of California Press.

Reeves, Jeffrey. 2013. "Sino-Mongolian Relations and Mongolia's Non-Traditional Security." *Central Asian Survey* 32 (2): 175–88.

Robbins, Joel. 2004. *Becoming Sinners: Christianity and Moral Torment in a Papua New Guinea Society.* Berkeley: University of California Press.

Robbins, Joel, and David Akin. 1999. "An Introduction to Melanesian Currencies: Agency, Identity, and Social Reproduction." In *Money and Modernity: State and Local Currencies in Melanesia,* edited by David Akin and Joel Robbins, 1–40. Pittsburgh: University of Pittsburgh Press.

Robinson, Kathryn M. 1986. *Stepchildren of Progress: The Political Economy of Development in an Indonesian Mining Town.* New York: SUNY Press.

Rohe, Randall. 1985. "Hydraulicking in the American West: The Development and Diffusion of a Mining Technique." *Montana: The Magazine of Western History* 35 (2): 18–29.

Rohrbough, Malcolm J. 1997. *Days of Gold: The California Gold Rush and the American Nation.* Berkeley: University of California Press.

Roitman, Janet. 2006. "The Ethics of Illegality in the Chad Basin." In *Law and Disorder in the Postcolony,* edited by Jean Comaroff and John L. Comaroff, 247–72. Chicago: University of Chicago Press.

Rolnick, Arthur J., and Warren E. Weber. 1997. "Money, Inflation and Output under Fiat and Commodity Standards." *Journal of Political Economy* 105 (6): 1308–21.

Romanov, Boris Aleksandrovich. (1928) 1952. *Russia in Manchuria (1892–1906).* Reprint, Ann Arbor, MI: J.W. Edwards.

Rosenberg, Daniel. 1981. "Leaders and Leadership Roles in a Mongolian Collective: Two Cases." *Mongolian Studies* 7:17–51.

Rossabi, Morris. 2005. *Modern Mongolia: From Khans to Commissars to Capitalists.* Berkeley: University of California Press.

Rozycki, William. 1996. *Concise Mongol-English and English-Mongol Dictionary.* Bloomington: C & M Press.

Rudnyckyj, Daromir. 2011. *Spiritual Economies: Islam, Globalization, and the Afterlife of Development.* Ithaca, NY: Cornell University Press.

Ruskin, John. 1862. *"Unto This Last": Four Essays on the First Principles of Political Economy.* London: Smith, Elder.

Sachs, Jeffrey D., and Andrew M. Warner. 2001. "The Curse of Natural Resources." *European Economic Review* 45 (4): 827–38.

Sahlins, Marshall. 1965. "On the Sociology of Primitive Exchange." In *The Relevance of Models for Social Anthropology,* edited by Michael Banton, 139–236. London: Routledge.

——. 1985. *Islands of History*. Chicago: University of Chicago Press.

——. 1992. "The Economics of Develop-Man in the Pacific." *Res* 21:13–25.

——. 1994. "Cosmologies of Capitalism: The Trans-Pacific Sector of the World System." In *Culture/Power/History: A Reader in Contemporary Social Theory*, edited by Nicholas B. Dirks, Geoff Eley, and Sherry B. Ortner, 412–55. Princeton: Princeton University Press.

Sallnow, Michael J. 1989. "Precious Metals in the Andean Moral Economy." In *Money and the Morality of Exchange*, edited by J. Parry and M. Bloch, 209–31. Cambridge: Cambridge University Press.

SAM (Support for Artisanal Mining). 2008. *A Sub-Program on Development of Artisanal and Small-Scale Mining until 2015: Appendix to the Governmental Resolution No. 71*. Ulaanbaatar, Mong.: Support for Artisanal Mining Project.

Samykina, E. V., A. V. Surkov, L. V. Eppelbaum, and S.V. Semenov. 2005. "Do Old Spoils Contain Large Amounts of Economically Valuable Minerals?" *Minerals Engineering* 18 (6): 643–45.

Sanders, Alan J. K. 1982. "Enter the Future, with a Deafening Roar." *Far Eastern Economic Review* 19 (November): 43–44.

Saunders, Richard. 2014. "Geologies of Power: Blood Diamonds, Security Politics and Zimbabwe's Troubled Transition." *Journal of Contemporary African Studies* 32 (3): 378–94.

Scott, James C. 1977. *The Moral Economy of the Peasant: Subsistence and Rebellion in Southeast Asia*. New Haven: Yale University Press.

Scott, Michael W. 2007. *The Severed Snake: Matrilineages, Making Place, and a Melanesian Christianity in Southeast Solomon Islands*. Durham, NC: Carolina Academic Press.

Scott, Rachelle. 2009. *Nirvana for Sale? Buddhism, Wealth, and the Dhammakāya Temple in Contemporary Thailand*. Albany: SUNY Press.

Shipton, Parker. 1989. *Bitter Money: Cultural Economy and Some African Meanings of Forbidden Commodities*. American Ethnological Society Monograph Series 1. Washington, DC: American Anthropological Association.

Simmel, Georg. (1907) 2004. *The Philosophy of Money*. Reprint, London: Routledge.

Simukov, Andrei D. 1933. "Hotoni" [Hotons]. *Sovrennaya Mongoliya* [Contemporary Mongolia] 3 (1): 19–32.

Smith, Adam. (1789) 1979. *An Inquiry into the Nature and Causes of the Wealth of Nations*. Reprint, Oxford: Clarendon Press.

——. 1978. *Lectures on Jurisprudence*. Edited by Ronald L. Meek, David D. Raphael, and Peter G. Stein. Oxford: Oxford University Press.

Sneath, David. 1993. "Social Relations, Networks and Social Organisation in Post-Socialist Rural Mongolia." *Nomadic Peoples* 33:193–207.

——. 2000. *Changing Inner Mongolia: Pastoral Mongolian Society and the Chinese State*. Oxford: Oxford University Press.

——. 2006. "Transacting and Enacting: Corruption, Obligation and the Use of Monies in Mongolia." *Ethnos* 71 (1): 89–112.

Sokolewicz, Zofia. 1977. "The Possibility of Individual Cultural Expression and the Values in Traditional Mongolian Pastoral Culture." *Ethnologia Polona* 3:47–59.

Sommer, Matthew. 2000. *Sex, Law, and Society in Late Imperial China*. Stanford: Stanford University Press.

Sraffa, Piero. 1960. *Production of Commodities by Means of Commodities: Prelude to a Critique of Economic Theory*. Cambridge: Cambridge University Press.

Ssorin-Chaikov, Nikolai. 2000. "Bear Skins and Macaroni: The Social Life of Things at the Margins of a Siberian State Collective." In *The Vanishing Rouble: Barter Networks and*

Non-Monetary Transactions in Post-Soviet Societies, edited by Paul Seabright, 345–61. Cambridge: Cambridge University Press.

Stemmet, Farouk. 1996. *The Golden Contradiction: A Marxist Theory of Gold, with Particular Reference to South Africa.* Aldershot, UK: Avebury.

Stewart, Kathleen. 1988. "Nostalgia—A Polemic." *Cultural Anthropology* 3 (3): 227–41.

Stewart, Pamela, and Andrew Strathern. 2002. *Remaking the World: Myth, Mining, and Ritual Change among the Duna of Papua New Guinea.* Smithsonian Series in Ethnographic Inquiry. Washington, DC: Smithsonian Institution Scholarly Press.

Strang, Veronica. 2004. "Poisoning the Rainbow: Mining, Pollution and Indigenous Cosmology in Far North Queensland." In *Mining and Indigenous Lifeworlds in Australia and Papua New Guinea*, edited by Alan Rumsey and James Weiner, 208–25. Oxford: Sean Kingston Publishing.

Strassler, Karen. 2009. "The Face of Money: Currency, Crisis, and Remediation in Post-Suharto Indonesia." *Cultural Anthropology* 24 (1): 68–103.

Strathern, Marilyn. 1988. *Gender of the Gift: Problems with Women and Problems with Society in Melanesia.* Berkeley: University of California Press.

———. 1992. "Qualified Value: The Perspective of Gift Exchange." In *Barter, Exchange and Value: An Anthropological Approach*, edited by Caroline Humphrey and Stephen Hugh-Jones, 169–91. Cambridge: Cambridge University Press.

———. 1997. "Double Standards." In *The Ethnography of Moralities*, edited by Signe Howell, 127–51. London: Routledge.

Sühbaatar, O. H. 2001. *Mongolyn tahilgat uul usny sangiin sudar orshvoi* [Sacred mountain and water sites of Mongolia]. Ulaanbaatar, Mong:: Mongolia Academy of Sciences Press.

Swancutt, Katherine. 2012. *Fortune and the Cursed: The Sliding Scale of Time in Mongolian Divination.* Oxford: Berghahn Books.

Szynkiewicz, Slawoj. 1977. "Kinship Groups in Modern Mongolia." *Ethnologia Polona* 3:31–45.

Tarde, Gabriel. (1985) 2012. *Monadology and Sociology.* Translated by Theo Lorenc. Reprint, Melbourne: re.press.

Taussig, Michael. 1980. *The Devil and Commodity Fetishism in South America.* Chapel Hill, NC: University of North Carolina Press.

Thomas, Nicholas. 1991. *Entangled Objects: Exchange, Material Culture, and Colonialism.* Cambridge, MA: Harvard University Press.

Thompson, E. P. 1971. "The Moral Economy of the English Crowd in the Eighteenth Century." *Past and Present* 50:76–136.

Tomlinson, Richard. 1998. "From Genghis Khan to Milton Friedman: Mongolia's Wild Ride to Capitalism." *Fortune*, December 7.

Trifonov, Ivan, and Yuri Krouchkin. 2000. *Mongolia: Its Mineral Resources and Law Encyclopedia.* Moscow: Admon.

Tseren, P. Buhan. 1996. "Traditional Pastoral Practice of the Oirat Mongols and Their Relationship with the Environment." *Inner Asia* 2:147–57.

Tsing, Anna L. 2015. *The Mushroom at the End of the World: On the Possibility of Life in Capitalist Ruins.* Princeton: Princeton University Press.

Tucci, Guiseppe. 1949. *Tibetan Painted Scrolls.* Vol. 1. Rome: La Libreria Dello Stato.

Tungalag, A., R. Tsolmon, and B. Bayartungalag. 2008. "Land Degradation Analysis in the Ongi River Basin." *International Archives of the Photogrammetry, Remote Sensing and Spatial Information Services* 37:1021–23.

Upton, Caroline. 2012. "Mining, Resistance and Pastoral Livelihoods in Contemporary Mongolia." In *Change in Democratic Mongolia: Social Relations, Health, Mobile Pastoralism and Mining*, edited by Julian Dierkes, 223–48. Leiden, Neth.: Brill.

Viveiros de Castro, Eduardo. 1998. "Cosmological Deixis and Amerindian Perspectivism." *Journal of the Royal Anthropological Institute*, n.s., 4 (3): 469–88.

Vreeland, Herbert Harold. 1954. *Mongol Community and Kinship Structure*. New Haven: Human Relations Area Files Press.

Wallace, Vesna A., ed. 2015. *Buddhism in Mongolian History, Culture, and Society*. Oxford: Oxford University Press.

Walsh, Andrew. 2003. "'Hot Money' and Daring Consumption in a Northern Malagasy Sapphire-Mining Town." *American Ethnologist* 30 (2): 290–305.

——. 2006. "'Nobody Has a Money Taboo': Situating Ethics in a Northern Malagasy Sapphire Mining Town." *Anthropology Today* 22 (4): 4–8.

Walsh, Margaret. 2005. *The American West: Visions and Revisions*. Cambridge: Cambridge University Press.

Watson, B. 1993. *The Lotus Sutra*. New York: Columbia University Press.

Weber, Max. 1904. *Die protestantische Ethik und der Geist des Kapitalismus* [The Protestant ethic and the spirit of capitalism]. Munich: CH Beck.

Weber-Fahr, Monika, J. Strongman, R. Kunanayagam, G. McMahon, and C. Sheldon. 2001. *Mining and Poverty Reduction*. Washington, DC: World Bank.

Werthmann, Katja. 2003. "Cowries, Gold and 'Bitter Money': Gold-Mining and Notions of Ill-Gotten Wealth in Burkina Faso." *Paideuma* 49:105–24.

Whitehead, Alfred North. 1920. *The Concept of Nature*. Cambridge: Cambridge University Press.

Whiteside, Kerry H. 2006. *Precautionary Politics: Principle and Practice in Confronting Environmental Risk*. Cambridge: Cambridge University Press.

Wolf, Eric R. 2010. *Europe and the People without History*. Berkeley: University of California Press.

Wood, Michael. 2004. "Places, Loss and Logging among the Kamula." *Asia Pacific Journal of Anthropology* 5 (3): 245–56.

Worden, Robert L., and Andrea Matles Savada. 1991. *Mongolia—A Country Study*. Washington, DC: US Government Printing Office.

World Bank. 2004. *Mongolia: Mining Sector Sources of Growth Study*. Washington, DC: World Bank.

——. 2006. *Mongolia: A Review of Environmental and Social Impacts in the Mining Sector*. Washington, DC: International Bank for Reconstruction and Development/World Bank.

Yakovleva, Natalia. 2007. "Perspectives on Female Participation in Artisanal and Small-Scale Mining: A Case Study of Birim North District of Ghana." *Resources Policy* 32 (1): 29–41.

Yenhu, T. 1996. "A Comparative Study of the Attitudes of the Peoples of Pastoral Areas of Inner Asia towards Their Environments." In *Culture and Environment in Inner Asia*. Vol. 2, *Society and Culture*, edited by Caroline Humphrey and David Sneath, 3–24. Cambridge: White Horse Press.

Yeshe, Lama, and Lama Zopa Rinpoche. 2012. *Wisdom Energy: Basic Buddhist Teachings*. Edited by Jonathan Landaw. Somerville, MA: Wisdom Publications.

Yule, Alexander D. 2007. "Changers: From Steppe to Market, and Beyond. Draft paper. http://alexyule.com/papers/ISP.pdf.

Zamora, Daniel, and Michael C. Behrent, eds. 2016. *Foucault and Neoliberalism*. Cambridge: Polity Press.

Zelizer, Viviana A. 1997. *The Social Meaning of Money: Pin Money, Paychecks, Poor Relief, and Other Currencies*. Princeton: Princeton University Press.

——. 2004. "Circuits of Commerce." *In Self, Social Structure, and Beliefs: Explorations in Sociology*, edited by Jeffrey Alexander, 122–44. Berkeley: University of California Press.

Zhengyin, L. 2003. "Arts of the Book, Painting and Calligraphy. Part 3: Eastern Central Asia." In *History of Civilizations of Central Asia*. Vol. 5, *Development in Contrast: from the Sixteenth to the Mid-Nineteenth Century*, edited by I. Iskender-Mochiri, 610–28. Paris: UNESCO Publishing.

Zimmermann, Astrid E. 2011. "Enacting the State in Mongolia: An Ethnographic Study of Community, Competition and 'Corruption' in Postsocialist Provincial State Institutions." PhD diss., University of Cambridge.

——. 2012. "Local Leaders between Obligation and Corruption: State Workplaces, the Discourse of 'Moral Decay,' and 'Eating Money' in the Mongolian Province." In *Change in Democratic Mongolia: Social Relations, Health, Mobile Pastoralism, and Mining*, edited by Julian Dierkes, 83–109. Leiden, Neth.: Brill.

Znoj, Heinzpeter. 1998. "Hot Money and War Debts: Transactional Regimes in Southwest Sumatra." *Comparative Study of Society and History* 40 (2): 193–222.

INDEX

CPSIA information can be obtained
at www.ICGtesting.com
Printed in the USA
BVOW03s1600260317
479447BV00001BA/19/P

9 781501 707551